Will Martindal

Responsil
Investment

An Insider's Account of What's
Working, What's Not and Where Next

Will Martindale
Canbury
London, UK

ISBN 978-3-031-44535-4   ISBN 978-3-031-44536-1   (eBook)
https://doi.org/10.1007/978-3-031-44536-1

Cover illustration: echo3005 and ImageFlow (2 images)

This Palgrave Macmillan imprint is published by the registered company Springer Nature Switzerland AG
The registered company address is: Gewerbestrasse 11, 6330 Cham, Switzerland

Paper in this product is recyclable.

Acknowledgements

Thank you to Elodie Feller and Rory Sullivan, who, for more than a decade, have helped me navigate the world of responsible investment.

To Rob Nash and Emmet McNamee who generously provided feedback throughout.

To Ben Wilmot, Olivia Mooney, Morgan Slebos, Paul Chandler, Margarita Pirovska, Alyssa Heath, Keith Guthrie, Dennis van der Putten and Karin Pasha for their insights throughout our time working together.

And to Fiona Reynolds and Nathan Fabian, who supported me throughout my time at PRI, and to all my colleagues at Cardano and NOW: Pensions.

To those I interviewed for this book: Alex Edmans, Bob Eccles, Catherine Howarth, Claudia Chapman, David Blood, Eelco van der Enden, Erinch Sahan, Fiona Reynolds, Jon Lukomnik, Martin Spolc, Nathan Fabian, Nick Robins, Philippe Zaouati, Richard Roberts, Roger Urwin, Sofia Bartholdy and Stephanie Pfeifer.

Contents

About the Author

Will Martindale is co-founder and managing director at Canbury. Using the latest technologies, Canbury provides investors, companies and NGOs with decision-useful sustainability data sets, experienced advice, research and stewardship.

Previously, Will was head of sustainability at Cardano and NOW: Pensions, responsible for specialist input on all aspects of Cardano's sustainability activities. Will also co-chaired IIGCC's policy advisory group and participated in PRI's legal framework for impact project and stewardship resourcing group.

From 2013 to 2020, Will was director of policy and research at PRI, leading PRI's global regulatory affairs and public policy activities. Will has a background in banking, joining JPMorgan's graduate programme in June 2004.

Will holds an M.Sc. in Comparative Politics from the London School of Economics and a B.Sc. in Maths from King's College London.

1

The Start

Why RI

I studied maths. Like many other maths grads, I started my career in finance. "Only for a year or two," I told myself, "to pay off my student loan." I joined JPMorgan's graduate scheme in the summer of 2004. In 2006, my role moved to New York. A year later, it was the start of the global financial crisis.

JPMorgan announced its takeover of Bear Stearns on a Sunday in Spring 2008. I was at home when I heard the news.

This was pre-smartphone. I received a text telling me to come into the office, a 30-something story office block at 270 Park Avenue. I worked in credit derivatives. My specialism was managed synthetic CDOs. I'll explain what these are later.

My first job was to price the trades between JPMorgan and Bear Stearns. Of course, JPMorgan and Bear Stearns traded with many investment banks, but by starting with the trades between both banks we could test Bear Stearns' risk management systems. Theoretically, pricing should be equal and offsetting. If JPMorgan's up 5 million USDs, Bear Stearns should be down 5 million USDs. The derivative market is a zero sum game.

But of course they were anything but. I was 25 and this was my first job out of university. I trusted the models. They were complex enough. But over the next few days I realised they were finger in the air. The value of these multi-million dollar CDOs (and we're talking hundreds of millions of dollars) was anyone's guess.

I didn't understand the point of it all.

© The Author(s), under exclusive license to Springer Nature
Switzerland AG 2023
W. Martindale, *Responsible Investment*, https://doi.org/10.1007/978-3-031-44536-1_1

And so, by Spring 2010 I'd left JPMorgan. Unsure where to take my career, I worked for the UK Labour Party in East London, a small charity on the Rwanda Congo border called Rwanda Aid, and the French bank BNP Paribas, before I ended up at the British charity Oxfam in their private sector team.

My finance colleagues were brilliant. The top of their class. I loved the job and worked with some great people.

During the financial crisis, I remember playing liars poker with dollar bills when the markets were quiet. I have a first-class honours degree in Maths. But not once did I outsmart my colleagues.

But I felt that my role lacked purpose. That I needed to do something more. I remember handing in my resignation at BNP Paribas. When I shared my news, first my manager, then a more senior manager and then an even more senior manager asked me why I was leaving. To my managers, it didn't make sense. But to me, it did.

I wanted to bring together my skill-set and social values, to work on projects that attempted to bring private capital to sustainability goals, and on issues such as climate change.

Oxfam wasn't quite as fun as I'd hoped. Many Oxfam colleagues viewed the private sector team with suspicion: public sector good, private sector bad. Engaging the private sector, and in my case, investors and banks, was to get Oxfam's hands dirty. I spent quite a bit of time justifying my role to colleagues.

I spent just over a year at Oxfam (Oxfam faced funding issues so my time there was short). But while at Oxfam, I discovered UN PRI, the United Nations-supported Principles for Responsible Investment.

At Oxfam, the private sector team was working on a project called Behind the Brands. We'd assessed six food and beverage companies, well-known brands such as Kellogg's, Mars and Unilever, on a series of sustainability themes. We met with PRI staff at a cafe in Shoreditch, East London, to discuss our findings.

The PRI was new to me, but it was exactly what I was looking for. The PRI offered a vision for how investors could steward private companies to achieve sustainability objectives. I was impressed with the approach and the calibre of the staff.

At the time I was also Labour's MP candidate for Battersea, South West London. This was unpaid and I'll talk a bit more about this later too. I applied for a role in public policy at the PRI and got it. I think my interest in politics helped.

After seven years at the PRI, the world's leading responsible investment group and at responsible investor Cardano, I've had a privileged front-row seat to the growth of an industry: Responsible investment.

And, oh my, what growth. When I joined the PRI in 2014, there were around 30 PRI employees. When a colleague had an afternoon off for a doctor's appointment, they'd email the whole staff. When I left PRI in 2020 there were over 200 employees. Revenue was £20 million or so a year. And there were 4,000 investors signed up to the PRI's six Principles.

PRI is part NGO, part service provider. In 2014, it was transitioning from former to latter under the leadership of Fiona Reynolds.

Reynolds took over from PRI's founder, James Gifford, who had interned with UN Environment Programme Finance Initiative, and went on to be PRI's managing director. Gifford had the charisma, determination, intelligence and passion to establish the Principles.

Gifford was liked and respected, but PRI was at an inflection point (twas ever thus?) and Reynolds was the experienced safe pair-of-hands that would help professionalise PRI.

As just one example of the changes underway, a colleague who joined a couple of years earlier described with wonder when Reynolds booked a three-star hotel for a business trip. Previously, PRI staffers would book a bunk in a youth hostel. Reynolds was at the helm for nine years of remarkable growth.

While a contributor to the growth in responsible investment, PRI was not solely responsible. Responsible investment was an idea very much of its time.

So much so, that in November 2021 I found myself walking the streets of Glasgow for the UK-hosted COP 26, the make-or-break climate conference. Nearly every suited and booted private sector delegate had ESG in their title. "ESG" was so mainstream it was borderline banal. Everyone works in ESG these days.

But for all the noise, as responsible investment has matured, both its potential and its limitations have become more apparent.

I have worked in responsible investment for over a decade, and I have seen and been part of its growth. I've also seen its limitations. My reflection is that responsible investment is profoundly important, but if you define responsible investment to include real-world sustainability impact, it remains unproven.

In other words, with few exceptions, responsible investment is not contributing to real-world sustainability impact at anything like the scale necessary to make a difference.

Part textbook, part briefing, part story-telling, this book is about the growth of an industry, what it's got right and what it's got wrong.

I set out how we got to where we are and—if responsible investment is to matter—where we go next.

Outline

The content of the book refers to notes I've taken throughout my career in responsible investment. I wouldn't call it a diary as such. But I am a note-taker, notes which I've saved and organised.

I've also interviewed a number of individuals in order to check my recollection of events, my interpretation of regulation or an initiative, and to test some of my ideas about what comes next.

The book is loosely structured in two ways. By chronology and by theme. But rather than be strict with my structure, I've done my best to favour readability.

My intention is to put forward a short, accessible and personal interpretation of what responsible investment is, what's working, what's not and where next. Perhaps at heart, I'm a campaigner. And I want responsible investment to be far more impactful than it currently is. That's my motivation for writing this book.

I expect the book to be of the most immediate value to responsible investment professionals or aspiring responsible investment professionals. I've done my best to explain the terminology. But it does assume some knowledge of the concepts and themes.

It is also about how new ideas can emerge, can gain traction and evolve to maturity, as well as the challenges and the barriers along the way.

Responsible investment is going through a period of rapid change. Responsible investment is evolving from being industry-led and voluntary to being mandatory. There's a growing demand for responsible investment from savers. Regulators are working to tackle greenwashing. There's a proliferation of new sustainability topics.

Responsible investment is also "growing up." And part of growing up is acknowledging limits, complexities and making difficult decisions.

Responsible investment has got this far often by trying to stick to the script, safety in numbers, and not talking about the awkward bits in public.

But this is no longer an option, and so it's not surprising, that it draws its critics.

Responsible investment needs to be clear about its capabilities, objectives and limitations. A challenge responsible investment professionals are rising to.

I've started the book with an overview of responsible investment's early history as well as prevailing definitions of responsible investment terminology.

This includes an overview of fiduciary duties, ESG integration, stewardship and financial materiality.

As responsible investment evolved, so too has responsible investment regulation and the many stakeholder groups working on responsible investment topics. I include a selection here.

I've set out the arguments put forward by those that criticise responsible investment, in particular, an increasingly vocal group of US Republicans.

Next, I've explored climate change in more detail, COPs 21 and 26 and climate change-related metrics, as well as a range of issues that come under the broad umbrella of responsible investment, including biodiversity, human rights and water.

I've provided an overview of European leadership on responsible investment, but also the challenges with its centrepiece regulatory intervention, the Sustainable Finance Disclosure Regulation (SFDR), as well as the UK's Sustainability Disclosure Requirements (SDR).

The final chapters look at corporate disclosure, impact investment, another look at stewardship, and I end with systems change.

But before we get to that, I'll start with the basics.

Terminology

Inevitably, we must start with terminology. ESG is more an adjective than a noun. ESG integration. ESG incorporation. ESG issues. ESG factors. At a push, ESG investment.

ESG, confusingly, is used to describe both investment products and investment processes. In my view, it should be reserved for the latter.

When used for investment processes, ESG tends to be a term associated with financially material risks. It represents a practice, whereby, investors consider (and the word "consider" is subjective; there exists a wide range of interpretation here) environmental, social and corporate governance issues in their investment decisions. Investors assess the ESG issues the company is exposed to, how it manages those ESG issues, and whether the exposure and management are incorporated in the price.

There are of course companies more exposed to some ESG risks than others: A mining company for example, aviation, cement or apparel. And different ESG risks vary in financial materiality.

As such, "ESG investing" is, well, investing. All investing should consider all material risks, regardless of their origin. Perhaps ESG investing means more regard than typical for ESG issues. But the motivation for considering ESG issues is the financial performance of the investment. Nowadays, integrating ESG issues is considered just part of an investor's job.

ESG is also used to describe investment products. ESG products are not adequately regulated (in other words, there is no such thing as an "ESG product"). If I had to generalise, an ESG product is one where the investor seeks to achieve some sort of "ESG outcome", that could, for example, be a portfolio that invests in higher performing ESG-rated companies or a portfolio that pursues an environmental target, such as net zero greenhouse gas (GHG) emissions. We'll look at ratings, net zero and sustainability themes later.

Responsible investment is the catch-all term that includes all of the above.

The terms responsible and sustainable tend to be used interchangeably. For some investors, sustainability may refer to a higher bar (in other words, sustainable investment is more sustainable than responsible investment). Sustainability is the term used by the UN (in the Sustainable Development Goals) and NGOs.

Sustainability is more widely understood than "ESG".

The UN Bruntland Commission defined sustainability as "meeting the needs of the present without compromising the ability of future generations to meet their own needs" (United Nations, 1987).

Or in other words, sustainability means lifting a billion people out of poverty and doing so within planetary boundaries. As such, the implication is that sustainable investment can contribute to this.

If I'm talking to clients I tend to use "sustainable", because it's a more intuitive term. Indeed "responsible" (and perhaps, sustainable too) is a normative term, even if it is bounded by certain assumptions. To some extent, advocates of responsible investment are working to shift the normative position on what constitutes "responsible".

But for the purposes of this book, I will use responsible.

Another term is SRI. SRI stands for socially responsible investment. Sometimes, SRI stands for sustainable and responsible investment, but the former is more common. SRI tends to be associated with exclusions, and includes an explicit "ethical" dimension, for example, excluding companies on religious or moral grounds, such as gambling and tobacco companies.

Another approach is what's called "best in class", where, rather than a negative screen, the investor applies a positive screen. The investor may still invest in tobacco companies but, for example, only invest in the companies that have strong policies in place to eradicate instances of child labour in supply chains.

Active ownership and stewardship (the terms are used interchangeably), which comprises of engagement and voting, are the ongoing management of an investor's assets. For equity (or stocks, or shares, also terms used interchangeably) it includes the right to attend, participate in and vote at company AGMs, as well as file (or, if in collaboration with others, co-file) resolutions to be discussed and voted on at company AGMs.

The word sustainable is also used to describe an investment where there is a specific sustainability objective. Green or environmental is a flavour of sustainability. Green bonds, for example, are similar to traditional bonds, in that they have a coupon or interest rate, a notional and a maturity, but the use of proceeds (what the bond is financing) must achieve an environmental objective. The Green Bond Principles help companies and investors establish what constitutes green (ICMA, 2021).

Blue bonds relate to water, both limiting the use of water in water scarce regions, addressing water pollution and the preservation of oceans from, for example, over-fishing. The word brown is sometimes used to describe a bond issued by a polluting company, although I prefer "grey" or "traditional".

TCFD stands for the Task Force on Climate-related Financial Disclosures and it is a framework for disclosing financially material climate change-related risks and opportunities.

Net zero greenhouse gas emissions means not adding to the amount of GHG emissions. GHGs include carbon dioxide, methane, nitrous oxide and flourinated gases, although we tend to just say "carbon", as in, "carbon footprint". For ease of measurement, we standardise all GHGs into a carbon dioxide equivalent, or CO2e.

In the world of responsible investment, typically "net zero" is short hand for "net zero greenhouse gas emissions by 2050", which is what's necessary if we're to achieve the objective of the Paris Climate Agreement, limiting global warming to 1.5 degrees Centigrade.

Still today, nearly every conference I attend includes a question on terminology. My stock answer is not to worry about it. In any nascent industry, and largely unregulated industry, terminology evolves.

If the SEC was to intervene with clear definitions of what we mean by each term that would, I think, be welcome. But in the meantime, it shouldn't stop responsible investment.

Intermediation Chain

It's also worth taking a moment to consider the investment or "intermediation" chain.

I find that responsible investment professionals tend to "stay in their lane", thinking about sustainability from the perspective of their employer, not how various parts of the intermediation chain interact, and how therefore we can be successful in achieving real-world sustainability impact by influencing and working with the intermediation chain around us.

While location-specific and evolving, the investment chain tends to be structured as follows.

At the top, are asset owners. Asset owners include pension funds ("schemes" or "plans"), insurers, endowments and foundations. The words "fund", "scheme" and "plan" tend to be used interchangeably.

Pension funds vary in size and structure. The Australian and Dutch pension fund markets are characterised as well-resourced, well-governed and well-funded. Funds such as ABP or PGGM in the Netherlands or HESTA or Cbus in Australia, are multi-tens, or in ABP's case, multi-hundreds, of billions of Euros or dollars of assets under management, where investment decision-making is often in-house. In other words, the pension funds themselves invest, or, in ABP's case, have their own dedicated asset manager. They may outsource investment to an asset manager for some asset classes (typically, more specialised investments, such as private equity, but the rest is done in-house).

The US, UK and Canadian pension markets have some well-resourced, well-governed and well-funded pension funds. In California, CalPERs, the public sector pension fund and CalSTRS, the pension fund for teachers, are two of the largest pension funds in the world.

The US, UK and Canadian pension markets also have a "long tail", thousands of smaller pension funds with a few billion, a few hundred million or even a few tens of million dollars or pounds in assets under management. For the smaller schemes, and even some of the bigger schemes, like USS, the UK's Universities Superannuation Scheme, with 400,000 members and around 90 billion pounds in assets under management, the pension fund will outsource some of its investment decisions to external asset managers.

Asset managers are intermediaries. Asset management is highly competitive, but can also be highly lucrative. Asset owners pay asset managers a fee to invest their assets. Fees vary considerably. I've seen some passive funds (a passive fund is a fund that tracks an index, like the S&P 500) with fees as low as 1 or 2 basis points, or 0.02% of assets under management, or in other words, 200 dollars per 1 million dollars invested.

But the more complex or successful investment strategies can charge as much as 1 or 2% of assets under management, as well as a performance fee of as much as 20% of positive performance. This can include hedge funds, private equity, infrastructure or property investments, as well as active funds (an active fund is a fund where the asset manager selects the companies based on their expertise and research).

Service providers support asset managers and asset owners by providing data, research and advice. For responsible investors, this includes ESG scores and research, provided by companies such as MSCI or Sustainalytics. There is an entire industry dedicated to ESG data, with providers competing on their ability to provide investors with high quality, timely data from a range of sources, in a format that investors can integrate into their investment processes.

A growing, but still niche area of service providers support asset managers and asset owners with engagement and voting. Companies such as Hermes EOS, Sustainalytics (again), ISS and Glass Lewis undertake engagement of companies or voting at companies' AGMs on behalf of asset owners and managers.

In common law countries, like the US or UK, asset owners tend to be governed by trustees with a fiduciary duty to their savers. In other words, the trustee must make investment decisions in the best interests of savers in the scheme. This isn't easy. Pension schemes may have millions of savers. "Best interests" tends to be interpreted as best financial interests.

In the US and UK pension fund trustees must seek independent advice, typically from an investment consultant (for example, Mercer or Willis Towers Watson). I would consider the investment consultant a service provider, however, consultants have evolved to provide fiduciary management.

In fiduciary management, trustees retain a fiduciary duty to pension savers, but delegate day-to-day investment decision-making across all the pension fund's assets to a fiduciary manager. In turn, the fiduciary manager may appoint external asset managers. In the US, this is called "outsourced CIO" (outsourced Chief Investment Officer).

We can think of the intermediation chain in five parts: savers, owners, managers, service providers and companies. Owners and managers do not achieve real-world sustainability impact in their own right, rather that's for the companies, supranationals or governments in which they invest. Owners and managers can however drive change within companies, supranationals or governments.

Each part of the investment chain is responsible for a different set of decisions and subject to a different set of regulations. It's no wonder terminology varies.

Materiality and Double Materiality

Another term that's important for responsible investment is "double materiality".

Before we understand double materiality, let's first turn to "materiality". An issue is financially material if it affects the value of the company, and therefore, the investment.

Take, for example, an apparel company reliant on cotton grown on land increasingly subject to drought due to climate change. Issues such as climate change, water use and water scarcity are financially material to the company. (So too, are many non-sustainability-related issues financially material.)

But as responsible investment has evolved so too has the approach. Investors have increasingly adopted double materiality as a feature of their investment strategy. The word double refers to the impacts of investments or company activities on the real world.

If we think about this in terms of objectives, the objectives are:

- Optimise risk-adjusted returns (which includes the integration of financially material ESG issues).
- Optimise real-world sustainability impact (or minimise negative real-world sustainability impact).

The term real-world is a bit clunky but basically we mean something that happens or changes in the real world (in our day-to-day lives) as a result of an investment decision.

All investments have real-world impact. For investors that adopt double materiality objectives, we're being clear on intentionality. Whether the investor is intentionally contributing to real-world impact and whether that impact is positive (or minimises negative) real-world sustainability impact (for example, reducing or even replenishing water use in the manufacturing of clothing).

Some investments may be attractive from a risk-adjusted return perspective, but cause too negative a real-world impact, and so be excluded. Other investments may be attractive from a real-world impact perspective, but not provide competitive financial returns.

There are various flavours here too. Some investors approach real-world impact from first principles, where the primary motivation is impact. UK law firm Freshfields labels this "ultimate ends" investing for sustainability impact. Others would call this concessionary.

Here, the investment product has a specific impact objective, set out in the terms of the investment, that may trump a financial objective.

While this may be desirable for some private investors, it is unlikely to become the prevailing approach.

Other investors approach real-world impact by considering risks (in a more expansive interpretation of risk where the risks are long term or systemic). Freshfields calls this "instrumental" investing for sustainability impact.

The lens remains financial performance, but for most types of investors, it's not surprising that the investment must be justified based on financial performance.

Indeed, this still requires "will". Investors may lack empirical evidence that proves financial materiality for a particular sustainability issue. We tend to assess future investment performance by considering historical returns, which are, by definition, backward-looking whereas many sustainability issues are inherently forward-looking.

Even if there is empirical evidence, it is rarely incontrovertible, and incentives in the market are not strong enough to overcome this inertia.

But for some investments, there may be a trade-off between issue and performance even over the long-term. In this case, I'm comfortable with responsible investors saying, "it's not for us, we're focusing where we can have impact, not where we can't."

There are also significant limiting factors on investors' agency, beyond trade-offs. Often, investors are too remote from some issues to be effective agents to consistently address them.

Double materiality objectives should be defined in a way that allows for a clear objective, a stated theory of change and a set of processes that allow the investor to make progress towards the objective.

There are two other sets of terms that are important when we think about double materiality: SDGs and Taxonomies.

The Sustainable Development Goals (UN, 2015), or SDGs, are the UN's framework for sustainable development, labelled a "shared blueprint for peace and prosperity for people and the planet, now and into the future."

Taxonomies are classification tools to help investors determine whether an economic activity is consistent with a public policy goal. In the case of the EU Taxonomy, the public policy goal is the Paris Climate Agreement.

In 2021, the European Commission defined responsible investment as "a comprehensive approach which consists of the systematic integration of both financially material sustainability risks (outside in) and sustainability impacts (inside out) in financial decision-making processes. It is crucial that both angles of the materiality concept are duly integrated for the financial sector to contribute pro-actively and fully to the success of the European Green Deal" (European Commission, 2021).

Here, the European Commission adds real-world sustainability impact ("inside out") to its definition of responsible investment.

To place materiality and double materiality, it is helpful to refer to a typography, developed by specialist sustainability investor, Bridges Ventures, and the impact coalition, the Impact Management Project (IMP).

It was first published in 2012 and updated in 2015 (by Bridges Ventures, the PRI, and the UK Impact Investing Institute among others), however its core is unchanged and it is often cited by investors and academics.

It sets out five approaches to investment:

1. Traditional: Limited or no consideration of ESG issues.
2. Screened: Negative or positive screened investments based on ESG criteria.
3. ESG integration: The consideration of ESG issues in investment decision-making.
4. Themed: The explicit consideration of ESG themes, such as climate change, clean energy, water use or biodiversity, in portfolio construction.
5. Impact: Investments with an explicit objective to achieve real-world impact. This category often divides into impact, with market returns, and impact, with some financial trade-off.

Today, ESG integration is a requirement of fiduciary duties, and so what was labelled "traditional" falls away and impact is typically at competitive risk-adjusted returns, again to be consistent with fiduciary duties. Sometimes, there is a sixth approach, philanthropy, but I would not consider philanthropy a form of responsible investment.

"Risk-adjusted" is important wording. Comparing just "returns" is somewhat subjective, as it is returns per unit of risk, and the investment's contribution to a portfolio's strategic asset allocation.

References

BBC, What is climate change? A really simple guide. [online]. Available from: https://www.bbc.com/news/science-environment-24021772 (Accessed, January 2023).

Cardano (2022), Sustainable Investment Policy. [online]. Available from: https://www.cardano.co.uk/wp-content/uploads/sites/3/2022/10/Cardano-ACTIAM-Sustainability-Policy.pdf (Accessed, January 2023).

CBC, 595 people were killed by heat. [online]. Available from: https://www.cbc.ca/news/canada/british-columbia/bc-heat-dome-sudden-deaths-revised-2021-1.6232758 (Accessed, January 2023).

Freshfields (2021), A legal framework for impact. [online]. Available from: https://www.freshfields.com/en-gb/our-thinking/campaigns/a-legal-framework-for-impact/ (Accessed, January 2023).

FSB (2023), Task Force for Climate-related Financial Disclosures. [online]. Available from: https://www.fsb-tcfd.org/ (Accessed, January 2023).

Gates (2021), How to avoid a climate disaster.

ICMA (2021), Green Bond Principles. [online]. Available from: https://www.icmagroup.org/sustainable-finance/the-principles-guidelines-and-handbooks/green-bond-principles-gbp/ (Accessed, January 2023).

IPCC, Climate change 2022: Impacts, adaptation and vulnerability. [online]. Available from: https://www.ipcc.ch/report/sixth-assessment-report-working-group-ii/ (Accessed, January 2023).

PRI (2015), What is Responsible Investment? [online]. Available from: https://www.unpri.org/an-introduction-to-responsible-investment/what-is-responsible-investment/4780.article (Accessed, January 2023).

PRI (2020), Investing with SDG outcomes. [online] Available from: https://www.unpri.org/sustainable-development-goals/investing-with-sdg-outcomes-a-five-part-framework/5895.article (Accessed, January 2023).

UNEP FI (2005), A legal framework for the integration of environmental, social and governance issues into institutional investment. [online]. Available from: https://www.unepfi.org/fileadmin/documents/freshfields_legal_resp_20051123.pdf (Accessed, January 2023).

United Nations (1987), Bruntland Commission. [online] Available from: https://sustainabledevelopment.un.org/content/documents/5987our-common-future.pdf (Accessed, January 2023).

United Nations (2015), Sustainable Development Goals. [online]. Available from: https://sdgs.un.org/goals (Accessed, January 2023).

2

Definitions of Responsible Investment

In Their Own Words

In research for this book, I asked a number of leading responsible investment professionals and commentators for their definitions of responsible investment. These are all individuals who I admire and who have had an effect on the way I think about responsible investment.

While there were a range of views across the interviews, my takeaways from the interviews were:

1. Responsible investment goes beyond ESG integration. This is now a well-trodden observation, but it's the almost casual acceptance that struck me as important from senior responsible investment professionals that for investment to be responsible, it must do something beyond integrate ESG issues.
2. Through investment decision-making and engagement with companies, regulators and stakeholders, responsible investment seeks to achieve some sort of positive real-world change or outcome.
3. The motivation for doing so may be financial, for example, issues that represent a systemic risk or issues that threaten a company or sector's social licence to operate. The motivation may also be intrinsic, based on values and ethics.

Here's what they said.

Bob Eccles, Professor of Management Practice at the Harvard Business School, said "I see responsible investment as grounded in the early days of

W. Martindale, *Responsible Investment*, https://doi.org/10.1007/978-3-031-44536-1_2

SRI which had a strong values-based approach. I go with the meaning that it once had, thinking about the values, rather than value."

"There are critical differences between ESG integration and impact. This is one of the problems you get into with ESG funds. They said they were going to resolve climate change. That's ridiculous. ESG factors are like any other type of risks. That's operations and activities."

"While there can be upside to ESG the real upside is in a company's products and services—their positive and negative externalities. Material risk factors and impact are analytically distinct. A company can be good or bad on one or the other or both."

For Jon Lukomnik, author, academic and managing director at Sinclair Capital, it's more about the distinction between responsible investment and impact investment.

"All investments have impact. So perhaps we could say that responsible investment is an awareness of the impact that your investments have, seeking to mitigate negative impacts and accentuate positive impacts within your particular risk return profiles."

"Impact investing is investing with the intentionality to have a specific impact. I don't think you have to go that far to be responsible."

For Stephanie Pfeifer, CEO at IIGCC, it's "around managing your risks and opportunities but in a way that does not create systemic risk or has a negative impact on society and the environment."

"In terms of the contribution – there is huge potential – some of which we're already seeing."

"I think it's about ensuring that there is real-economy impact and this is done by engaging with all stakeholders from policymakers, regulators, corporates, civil society and investors – to encourage them all to pull in the same direction."

"Obviously, complete alignment between all these groups globally is challenging, if not impossible, but particularly on the policy front there is much more than can be done to help get more capital flowing towards investments that will better support climate change, whether that's mitigation, adaptation, resilience or nature."

Claudia Chapman, Head of Stewardship at the UK FRC, said "Integrating material ESG issues into investment decision-making is just taking a broader or modern set of financially material factors that happened to be categorised differently."

"When does investing become responsible? When either the allocation of capital or stewardship of that investment has the intention of having a positive influence or impact on the issuer, group of companies, sectors, environment, economy or society. It is when an investor uses their allocation decisions or influence to effect change, that as well as delivering financial return, improves the long-term prospects of that investment. I would include an investor actively exercising its voting rights in this too."

"Responsible investment also takes on issues of fairness and ethics that may not have a financial impact over the investment horizon. Addressing them may maintain that company's social licence to operate."

Nathan Fabian, Chief Responsible Investment Officer at PRI, said, "Responsible investment is knowing and doing something about the impact of your activities as an investor on your customers and stakeholders."

"By impact I mean, not just the legal constraints that are placed around the financial product or service, but the actual impacts observed and evidenced, and in today's world that means the impact of your investment activities on environmental sustainability goals that are taken up by governments and international rights frameworks."

When I asked about proportionality, Fabian added, "The framework I usually use is one of sustainability goal, alignment and performance. When you can understand material issues – such as a planetary boundaries framework or international agreements on rights – you have a point of calibration against which you can assess the performance of economic activities. As long as you are prepared to work with that, you can judge the proportionality of your responsibility."

Philippe Zaouati, Founder and CEO, of Mirova, said, "There are so many definitions of responsible investment. It's a lexicon jungle."

"The usual way to define responsible investment is to say that it's the integration of environmental and social impact in everything we do – it's a broad definition and it's a good definition."

"There are however a couple of very important differences between 'responsible investment' and 'investment'."

"The most important one is the different way you see the use of finance. Responsible investment is a proactive tool, whereas finance is usually seen as passive in that you finance the economy as it is. You wait for the economy to evolve and then you finance it."

"Responsible investment thinks differently. You can be ahead of the curve compared to the economy. Take a step beyond, be proactive, reallocate money in a proactive way to support your sustainability goals."

Steve Waygood, Chief Responsible Investment Officer, Aviva Investors, said, "[Responsible investment] considers ethical questions. It consciously thinks through the implications of portfolios today on future generations, bringing in sustainability."

"It thinks carefully about power and the influence of money, about how it can be deployed optimally to improve outcomes for all, ensures that the individual whose money it is can rest assured, in terms of how it's invested, and the voice on their behalf, is being used to drive, at least acceptable behaviour and ideally sustainability behaviour. It is about where conscience meets capitalism."

"Capital markets are amoral. Capitalism itself doesn't start with ethical considerations. Any investment deemed responsible needs to think these through."

"In isolation, integration of ESG issues so as to optimise alpha isn't responsible. It's the first step, but if it's the last step, then it isn't responsible. Responsible investment needs to include stewardship and at a minimum compliance with general ethical standards."

Nick Robins, Professor in Practice for Sustainable Finance at Granthan Research Institute, said, "There is no 'one' responsible investment. From the beginning, it's been a broad perspective about how apparently non-conventional factors can be important for the delivery of investment outcomes – and that gets us to ESG. I see ESG as ingredients or factors to deliver a range of often quite different investment outcomes."

"You can use ESG factors to better understand and manage financial risks and returns; you can use ESG factors also to deliver sustainability impact (in terms of reducing harm and contributing to social and environmental improvements) and you can use ESG factors to achieve alignment with societal objectives, such as net zero, ending poverty or the full set of Sustainable Development Goals."

"All of these are responsible investment and all will result in very different outcomes in terms of the assets you hold and the relationships you have with business and policymakers. This brings a positive creative tension, but it also brings confusion, along with greenwashing (or misbuying) as people find that what they had in their ESG funds was minimal materiality assessment rather than alignment. This fuzziness is perhaps one factor behind the ESG backlash underway in the US."

For Sofia Bartholdy, Net Zero Lead at the Church Commissioners for England investment division, responsible investment is "an umbrella term that covers different ways of including planetary and societal factors into the investment process with different objectives from ethical stances to risk management."

"Having ethical exclusions is a decision to avoid certain activities because of your values. ESG integration is considering risks and opportunities from environmental, social and governance factors, the same way you would consider management quality, cashflows, competitive advantage."

"ESG integration is neutral to the ethics of the issues and many fundamental investors will already consider some ESG factors, whether they label them or not."

"Seeking environmental or social outcomes is the third way of thinking about responsible investment. This includes everything from concessionary impact investing to investing in public markets either by finding companies that provide solutions to societal challenges, or where the investor believes that their stewardship activities can influence the investments to avoid harm or do good to people and planet."

I asked Bartholdy about the limitations of responsible investment.

"One of the fallacies with using RI as a term is that it is a very broad term and is often used to talk about something specific."

"ESG investing, ESG scores and ESG products, good ESG are meaningless as terms."

"Another key limitation is that, while I believe investors can be part of the solution, investors alone will not save the world."

"No matter which way you incorporate a responsible investment strategy, you should not, and most don't, forget the non-responsible part of the investment process."

Finally, I asked Alex Edmans, Professor of Finance at London Business School, about the evidence on responsible investment's contribution to real-world impact.

"A company that reduces its carbon footprint is clearly having a real-world impact, but the question is, can investors achieve real-world impact?"

"There are two main tools. One of those tools is stock selection: Divestment from so called brown companies and investment into green companies. The scope of that is relatively limited. Why? Because if you sell your stock in a company, you can only sell if somebody else buys."

"People will say, 'even if somebody else buys, you need to reduce the stock price in order to get somebody else to buy'. But there are lots of amoral investors out there – not immoral – just that their goal is purely financial return."

"If you sell out of fossil fuels, it might be that you're selling from companies with credible transition plans that they're putting into practice."

"Divestment is one channel where I think the impact is limited, particularly in the blunt way in which many investors do this."

"The second channel is engagement. I think engagement conceptually can add value, but sometimes engagement can go into micromanagement."

Has responsible investment promised too much? I asked.
Yes.

"I think this is why responsible investment has got a lot of push-back from the likes of Tariq Fancy."

"While I don't agree with the tone or the nature of Tariq Fancy's comments, I do think that the challenge is useful." [I will introduce Fancy's arguments later.]

"Responsible investment has promised too much."

"I support the idea of responsible investment, but I think the best people to use the tools, are the people who understand the limitations of those tools. You would like a Formula One racing driver to know how to use the brakes as well as the accelerator."

An Attempt at a Modern Definition

Here's how I define responsible investment.

Responsible investment is an approach to investment that intentionally maximises positive real-world impact and minimises negative real-world impact, through investment decision-making, stewardship and policy engagement, leading to sustainable benefits to the economy, the environment and society, while optimising risk-adjusted returns.

All investments have real-world impact. The impact might be small, possibly even inconsequential and certainly difficult to attribute, but it is nevertheless impact. The key for responsible investment is to understand that impact, maximise positive real-world impact and minimise negative real-world impact.

There are five characteristics necessary to understand how investors should approach real-world impact. This can be referred to as a Theory of Change. They are:

1. Goal: The real-world goal of the investment strategy.
2. Intention: How the investment strategy will achieve the goal.
3. Process: The processes the investment strategy will undertake.
4. Measurement and reporting: The metrics that will be used to measure real-world impact to savers and regulators.
5. Financial performance: The financial returns and risks of the investment strategy.

In aggregate, these five parts constitute the investor's additionality.

Additionality is the real-world sustainability impact that can be attributed to the investor (as opposed to, just the assets in the portfolio). This is important because investing in something "good" does not equate to positive impact unless the investor is, in some way, additive.

Additionality in its own right is hard to measure. That's because there's no base case. Perhaps, theoretically, I could run two investment strategies, one where I seek to demonstrate additionality, and one where I do not, and I compare the real-world impact of the two. In practice, that's not possible.

And so, rather than test additionality, I test each part of the five-part framework.

This means, that responsible investment, including real-world impact, can be regulated, because the regulator can provide parameters around each part of the five-part framework, requiring disclosure and scrutinising the disclosure. The regulation would address greenwashing improving confidence in responsible investment.

Investors have two primary channels: Investment and stewardship. Here, I'm incorporating policy engagement within stewardship (or perhaps more accurately, as an extension of stewardship). Investors apply the two channels in different ways depending on their investment strategy. The strategy is determined by the characteristics of the companies in the investment portfolio.

A company should be assessed in two ways, both the company's operations and the company's products.

Measuring a company's real-world impact is hard, given assessments are subjective, reflecting the complex nature of companies and the historical and geopolitical structures within which they operate (companies operating within a predominantly coal-dominated energy system will be more GHG emissions intensive than a renewables-dominated energy system).

Investors may use performance thresholds per activity or indicators such as capital expenditure on activities considered sustainable.

This should be combined with "do no significant harm" requirements. There are some companies with products that positively contribute to real-world impact but processes that cause significant harm.

Investment and stewardship are actions that investors already do, but to consider whether the investor is responsible, we can go back to the five-part framework. There may be investment strategies that do not set a goal, intention, process and measurement—but then, these investments wouldn't be considered responsible investments.

For me, responsible investment is at market rates of return, over time horizons consistent with the end investor (hence, optimising risk-adjusted returns in the above definition).

That's not to say that my own personal motivation is not intrinsic; it is. I work in responsible investment because I want to contribute to sustainability outcomes. But that responsible investment is, in most markets, regulated under risk and return parameters.

As soon as you introduce concessionary investing, the question becomes how much concession.

I also don't believe it's necessary. If the purpose is more sustainable capital markets, you can get there through addressing systemic risks, as this book sets out.

3

RI and PRI's Short Histories

In Brief

To understand the mishmash of terms, it's important to understand responsible investment's short history.

Responsible investment has its origins in ethics.

In the 1970s and 1980s, a handful of investors, often religious investors, such as the Church of England or the Quakers, or endowments of US universities, took steps to exclude companies from their portfolios that were doing business in South Africa. One such example was Barclays, eventually selling its South African subsidiary in 1986 under pressure from consumers, students and investors.

The boycott helped pave the way for responsible investment.

In 1992, the UN Environment Programme established its finance initiative. The topic, this time, was the environment. UNEP FI, as it became known, started by bringing together a group of bankers following the Earth Summit in Rio de Janeiro. The origins were modest, essentially to share information. But in some ways it, too, was groundbreaking, in that it paved the way for competitive, private sector companies, to work together on sustainability issues.

The Equator Principles, which cover environmental and social risks in project finance, were launched in Washington, D.C., in 2003.

Around the same time, investors started to follow suit, and in particular, asset owners (pension funds and insurance funds). In the early 2000s, UNEP FI, and a number of its investor members, such as the large Canadian pension plan, CPPIB, also started to discuss the idea of principles, which, I was

W. Martindale, *Responsible Investment*, https://doi.org/10.1007/978-3-031-44536-1_3

told in research for this book, were originally known as the Manhattan Principles. The principles, the asset owners hoped, would help guide investors on responsible investment.

The asset owners, supported by UNEP FI, set out six principles, including, to integrate ESG issues in investment decision-making. "ESG" was a new term for most investors.

The United Nations-supported Principles for Responsible Investment were launched by the then General Secretary of the United Nations, Kofi Annan, at the New York Stock Exchange in 2005. It was to be known as PRI.

The PRI's launch was, in part, thanks to legal analysis undertaken by UK law firm, Freshfields. The Freshfields' study (coined "Freshfields 1.0" by its lead author, Paul Watchman) asked, can or should, investors integrate ESG issues in investment decision-making, as part of their fiduciary duties? The answer was that, yes, they can, and arguably, they should. The study paved the way for other asset owners and their asset managers to sign up to the Principles.

While, in theory, conceptual clarity by a magic circle law firm should resolve investors' questions on responsible investment, in practice, it was anything but. The question of fiduciary duties persists till this day.

This is where this book starts in the early 2010s. My introduction to responsible investment was in 2011. But let's first turn to PRI.

The PRI

I will introduce a number of sustainable finance groups later, but the PRI is central both to my background and the global growth of responsible investment, so it makes sense to start with PRI.

The PRI was launched in 2006. It was the brainchild of a small group of individuals, with James Gifford, then an intern at UNEP FI, and Paul Clements-Hunt, UNEP FI's head, at its core. A group of pension plans committed to the Principles.

The first Principle is to integrate ESG issues in investment decision-making, the second is to be an active owner or steward, the third is to seek ESG disclosure, the fourth is to promote the Principles, the fifth is to collaborate and the sixth is to report.

Nowadays, there is nothing controversial here. In many markets, requirements similar to the Principles are codified in regulation. At the time though, it was significant, with "ESG" being the entry point for NGOs, the UN and sustainability professionals to the capital markets.

The PRI's mission, introduced a few years later, says that the PRI will work to address "obstacles to a sustainable financial system in market practice, structure and regulation" (PRI, updated 2021). The reference to public policy engagement, which is not explicitly included in the Principles, is particularly noteworthy, as it demonstrates how responsible investment evolved in its first few years, more explicitly considering the role of investors in broader systems change.

The PRI has three categories of signatory: Asset owners, asset managers and, what it calls, service providers, which is everyone else, including consultants and data providers. The PRI has around 5,000 signatories, although, the number of signatories tells only part of the story.

The PRI's governance model is skewed in favour of asset owners. Asset owners occupy more board positions and pay lower fees. This is by design. Asset owners sit at the top of the intermediation chain. Asset managers are their agents, acting on their behalf.

In practice, some asset owners are considerably less influential than their asset managers, to the extent that they are buying what they're sold. But in theory, asset owners are at the top.

This is certainly the case in Australia. The PRI has had three CEOs, James Gifford, Fiona Reynolds and David Atkin; all Australian. This is however not surprising for at least a couple of reasons.

Australia's superannuation market is well designed, well-funded and was at the forefront of responsible investment. The pool of investment professionals at CEO level with asset owner and responsible investment experience is not extensive and it's not surprising that many are therefore Australian.

And an Australian CEO means the PRI board does not have to pick between a European or a North American, each with its distinct approach to responsible investment.

In an interview for this book, I asked Fiona Reynolds about her appointment as CEO in 2013.

Reynolds said, "I was working at the Australian Institute of Superannuation Trustees (AIST) … [and] considered that Australia was a leader in ESG issues. We had for example been very early in working with superannuation funds to measure their carbon footprint and importantly to disclose that information."

"When the PRI was established, there were a number of Australians involved in the drafting group. The PRI was definitely something we as an industry wanted to get behind and get moving globally. It was needed. We certainty understood that you can't tackle these issues in isolation."

"I was at AIST when the PRI board were considering a new Managing Director role. We were just about to have another federal election in Australia and it was very clear that the labor government was going to lose."

"I was working on superannuation policy, and it was clear Australia was going to be heading backwards in both superannuation policy and on ESG issues so I decided it was a good time for a change and a move overseas seemed sensible and something I had always wanted to do."

"I didn't really think I would get this job. If you want to do something globally, you normally have to apply for a range of jobs and get to know the headhunters, so I thought that it was worth throwing my hat in the ring as a way of beginning my job search process."

"I applied and to my surprise was offered the role. I initially turned down the role but luckily someone talked sense into me, and the rest as they say, is history."

"ESG roles didn't pay very well back in those days and I was prepared to take the job for the money that was offered, which probably helped me get the role, as the PRI didn't have deep pockets, but I was very committed to the cause."

The PRI's engine room is its reporting and assessment framework, which PRI staff abbreviate to R&A and detractors somewhat disparagingly shorten to "survey".

All signatories must report. New joiners have a grace year. Increasingly investors want to use their regulatory disclosures instead of PRI reporting, with PRI incorporating regulatory disclosures such as TCFD, SFDR or stewardship code reporting.

The reporting is divided into modules depending on the investment strategy. Investors that invest in listed equity must disclose against the listed equity module and so forth. In 2023, this changed. Asset owners no longer had to disclose against the asset class modules.

Some questions are public, some are private. The PRI publishes a transparency report which includes answers to the public questions for all its signatories and an assessment report which scores signatories' answers. Each module has a score. The PRI puts the score into context providing a range of scores based on the type of signatory (asset owner or manager) and region. European asset managers, for example, are unsurprisingly scoring higher than, say, Asian asset managers, where responsible investment is a less well-established concept.

In the last few years, investors have been subject to a raft of voluntary and mandatory disclosure requirements. UK pension funds have published

TCFD reports, signatories to stewardship codes have published their stewardship and voting activities, European investors have disclosed against the Sustainable Finance Disclosure Regulation (SFDR) and begun to prepare Taxonomy reporting, signatories to the Net Zero Asset Managers Initiative (NZAMI) have disclosed their targets and their progress toward targets.

Investors, understandably, are concerned with "reporting overload." "And anyway," a CIO of a UK pension fund said to me, "no one reads the PRI reporting." Whereas, the UK's Financial Reporting Council (FRC) provides more bespoke feedback to signatories that have prepared UK stewardship code reporting, in the first instance at least, the PRI provides only "automated" feedback.

Regulatory reporting tends to include fund-level reporting. For large asset managers, entity-level reporting is not particularly useful for asset owners in their fund selection as their investment strategies vary considerably depending on the portfolio.

Challenges aside, a global responsible investment data set remains a worthwhile pursuit, and PRI is the only organisation that can deliver that.

I asked Reynolds about her first few years as CEO.

"When I started, there were a few priority objectives. The PRI didn't have a good governance structure. It had a council that was elected. It had a board that was appointed. It wasn't clear who was making decisions. It was unwieldy and many of the signatories didn't like it."

"There was a small executive, but the executive didn't have a lot of previous experience of running a membership-type organisation like PRI."

"I needed to get the governance right, which I worked on with the new Chair at the time Martin Skancke. Then I needed to get PRI to grow. At the time, it had very little revenue."

"The strategy was to grow the revenue and grow the services, and ultimately to mainstream responsible investment. I also wanted to improve accountability. I kept hearing that organisations signed up to the PRI, but did not take action, so it was clear we needed to expand and upgrade the PRI reporting and assessment framework".

PRI In Person is the PRI's flagship global conference. I expect this, too, will change, with investors more and more conscious of their own GHG emissions. But Reynolds' first PRI In Person is worth mentioning.

"About 6 months into my role, we had our annual conference, PRI In Person, in South Africa."

"At the AGM, there were lots of questions, particularly from the Danish signatories, around the governance structures and signatory voting rights. It was really quite difficult."

"The signatories were not happy with the answers they received from the board and around what actions they were proposing to take to improve board transparency and accountability."

"Following the AGM, there were a number of consultations with the Danish signatories to try to resolve issue, but in the end, a number decided to leave, citing concerns around governance. Happily they all came back a few years later."

I asked Reynolds if the governance issues at the time affected PRI's ambition on sustainability issues in the years ahead. I've long been of the view that during my time at PRI, the board was overly cautious about other signatories leaving and could have afforded to take more risk.

"No, I don't feel that the PRI lacked ambition, it just had to be realistic. In my view, we took a sensible approach with a global signatory base, with investors at different stages of their responsible investment commitments."

"Our job was to bring people along, and it's worked. That's why PRI is the leading global responsible investment organisation."

PRI has around 200 staff, with London as its headquarters. Its staff are divided into three broad areas:

- Operations staff, including reporting.
- Networks staff, which are based in region, and work directly with signatories.
- And content staff, which are mostly based in London, run working groups, organise webinars, prepare guides on process or theme and so forth.

I expect this too will change, with PRI more focused on delivering regional content.

One challenge for PRI is its breadth. PRI has worked on everything and anything, even looking into human rights at the 2022 men's football World Cup in Qatar. And so I also expect PRI's content programme will also become more focused.

After almost two decades, PRI has established considerable goodwill and cross-industry capital. My view is and always has been that for those working in responsible investment, PRI is something that belongs to all of us and that we should all support to succeed.

But for responsible investment's next chapter, my view is that PRI should evolve into advisory, thought leadership and global convening, with a focus on systemic stewardship.

For advisory, the PRI could start by providing personalised feedback on annual reporting. A next step could be to work more closely with a subsection of signatories providing hands-on help to support signatories implement their responsible investment objectives. This could involve, for example, direct secondments from PRI to signatories (and indeed, vice versa).

For thought leadership, signatories need to provide space for the PRI to prioritise topics PRI's board and staff consider important in meeting the SDGs—even if, they are not part of the current investor "to-do" list. There will be sustainability topics that are important, but not yet investable.

And PRI is and I think always will be responsible investment's global convenor. It has a unique opportunity to connect investors and policymakers across the globe. The human rights collaborative initiative, ADVANCE, is a current example. No other group would have PRI's global convening capabilities.

I asked Reynolds about PRI's future. "I felt when I left, PRI was becoming too slow in its decision-making and too bureaucratic, and it was starting to focus too much on itself as an organisation."

"You sit back as the CEO and say, 'how have I created this big bureaucracy, I am of course responsible?' I don't think Covid helped, I also think in part it was growing pains, PRI was getting quite large at this stage. There are now 5,000 signatories. How many of them are responsible investment leaders? A small percentage. It does not mean that there's not some great asset owners and managers, there are – just not enough of them."

"I don't think PRI's role is to do everything. It's best role, in my view is and has always been being a big tent organisation. It could of course, work more effectively in a regional way, working within countries, within their regulatory frameworks and recognising that different countries and regions are at different stages. That way it could work more effectively with the leaders and the laggards."

"PRI also has amazing convening powers, it can bring investors, regulators, policymakers and other stakeholders to the table. It can socialise the issues. But I don't think the PRI can do everything. Others, such as local groups or those who specialise in a particular issue, can then step up and run with issues either separately of in conjunction with the PRI."

And what of Reynolds' achievements?

"There are a few achievements I'm particularly proud of while at PRI. The overarching aim was to mainstream responsible investment. We did that. When I started at the PRI there was very little being done on climate change. We launched the Montreal pledge in 2014 and investors started to measure their carbon footprint, this then led to the creation of significant initiatives such as Climate Action 100+ and the Net Zero Asset Owners Alliance."

"We started to get S issues on the agenda, in particular, labour rights, human rights and to link social issues within climate change with a focus on a just transition."

A Brief Note on Climate Change

This is not a book on climate change. But, like many other responsible investment professionals, when school friends or neighbours ask what it is I actually do, I often say, "climate change." "Oh, you invest in wind turbines?" "Yes, something like that."

Invariably, this tends to be followed on their part by some sort of explanation of nuclear fusion or why we'll run out of lithium. Most empathise. "Someone has to work on climate change I guess."

I'm sure most readers will already have honed their explanations of what climate change is and why it matters.

For sceptical colleagues or clients (or indeed, school friends or neighbours) here's my explanation.

The problem is "tipping points and feedback loops."

Climate change refers to global warming caused by GHG emissions of human activity. This leads to the increased frequency and severity of weather events, such as droughts, sea-level rise, floods, heatwaves, hurricanes and wildfires.

Globally, we emit around 51 billion tons of GHG emissions a year. Most of our emissions come from industry (in particular cement, steel and plastic), energy (including electricity, heating and cooling), agriculture and transport. To stop climate change, we need to stop emitting GHG emissions.

Greenhouse gases trap energy from the sun in the Earth's atmosphere, warming the planet. We've already warmed the Earth to at least 1.1 degrees Centigrade (it's likely more than this, we measure temperatures as 10-year rolling averages) (IPCC 2022). The GHGs that trap energy in the atmosphere include carbon dioxide (CO_2), methane (CH_4), nitrous oxide (N_2O) and fluorinated gases.

However, temperature change is not uniform across the globe. The Earth is warming more rapidly at the Poles, by as much as 3 degrees Centigrade of warming. As the Poles warm, the ice melts. Ice is replaced with water, and while the ice reflects the sun's rays, water absorbs the sun's rays. This causes further warming.

The permafrost (ground that remains frozen) begins to thaw. Permafrost stores methane from hundreds of thousands of years of decayed animal and plant matter. Methane is a particularly potent greenhouse gas (multiples that of carbon dioxide). As the permafrost thaws, methane is released. This causes further warming.

As the Arctic warms further, there is less of a difference between polar air temperature and warmer equatorial air. This weakens the jet stream, which, typically acts as a barrier between the cold and warm air. This causes further warming at the Poles. It also causes extreme weather events (in 2021, British Columbia experienced a heat dome, with heat at 49.6 degrees Centigrade—much warmer than body temperatures—and as such, 595 people died) (and in December 2022, extreme cold across North America)(CBC 2021).

Carbon capture and storage does not work at scale. There are some attempts at technologies that suck carbon dioxide out of the atmosphere, but it always takes more energy to do that than the energy we gained that puts carbon in the atmosphere in the first place.

This is because, the concentration of atmospheric carbon dioxide, while the problem, is 420 parts per million. Even at source (say, alongside a coal-fired power station), carbon capture and storage is ineffective.

We could and should plant trees. However, planting trees can take 15–20 years for the trees to grow and sequester carbon (time we do not have). And the only way you sequester carbon emissions from forestry is to convert land that used to be something else into forestry.

The only action we have is to rapidly decarbonise energy systems and prepare for climate adaptation. On climate change at least, this is what responsible investment needs to do.

References

CBC (2021), 595 People Were Killed By Heat. [online]. Available from: https://www.cbc.ca/news/canada/british-columbia/bc-heat-dome-sudden-deaths-revised-2021-1.6232758 (Accessed, January 2023).
IPCC, Climate Change 2022: Impacts, Adaptation and Vulnerability. [online]. Available from: https://www.ipcc.ch/report/sixth-assessment-report-working-group-ii/ (Accessed, January 2023).

4

Do Your Duty

The UN

When I was about 13 or 14 I remember being asked to write down what I wanted to do when I grew up. I said I wanted to work at the UN.

The UN was a source of inspiration to me. My upbringing was ordinary (in a good way). I grew up in a suburban town 100 miles southwest of London and a short bike ride from the UK's South Coast.

I went to my local state school, a stone's throw from our new-build house on what, at the time, was one of Europe's largest housing estates, made up of thousands of other similar new-build houses.

The UN was anything but ordinary. It was foreign. It was exciting. It was where I wanted to work.

Shortly after I started working at PRI, I was invited to a meeting at UNEP FI in late Spring 2014. It was my first visit to Geneva. The city is beautiful. Nestled between mountains, it boasts a crystal clear lake, warm sunshine, snow-topped peaks and European cafes.

After work, we'd head to the riverbanks of the Rhône, floating downstream and walking upstream.

My first meeting was at the UN itself. I couldn't wait.

The UN building is known as the Palais. The building itself is not palatial, but there are beautiful views across Lake Geneva and the Jet D'Eau water fountain. On a clear day, you can see Mont Blanc.

Security guards waved me through a beeping metal detector, took a look inside my rucksack and ushered me towards the Palais and its maze of corridors.

© The Author(s), under exclusive license to Springer Nature Switzerland AG 2023
W. Martindale, *Responsible Investment*, https://doi.org/10.1007/978-3-031-44536-1_4

PRI and UNEP FI staff, along with a handful of investors, sat round a table putting together a budget for a project that sought to "end the debate on fiduciary duties," almost a decade after the publication of the first Freshfields report and the launch of PRI.

The meeting led to a major new project, Fiduciary Duty in the 21st Century.

In this section, I'll set out the next few years of responsible investment, with a focus on fiduciary duties, what they are and why they matter.

Fiduciary Duties

Admittedly, as a topic, fiduciary duties do not sound interesting. They are however important in understanding responsible investment and the growth of PRI, and so they warrant explanation.

Decisions made by fiduciaries cascade through the intermediation chain and in turn through the investment decision-making process, ownership practices, and ultimately, the way in which companies are managed.

Where there exists agency, where one party has agency over another, the relationship is often accompanied by fiduciary duties: A duty of care, loyalty and prudence, whereby the agent makes decisions in good faith and judgement in the best interests of the principal.

Take, for example, a dentist. The dentist has agency over the patient, known as the principal. The dentist will know better than the patient whether or not the patient should have preemptive dental treatment. However, the dentist may be paid to undertake that dental treatment. As such, the dentist has a duty of care to the patient and to act in the patient's best interests. If the dentist fails to do so, the dentist's licence would be revoked.

Similarly, investment decision-makers have a duty of care to their savers and to act in their savers' best interests. The agents, in the case of pension law, are known as trustees, a group of individuals tasked to make investment decisions on behalf of savers. The savers are the principal.

In most cases, the type of decisions made by the trustees include setting out investment beliefs, deciding in which assets to invest, and how much, selecting an adviser, selecting a lawyer and if the investments are not undertaken in house (by directly employed staff), selecting the asset manager or managers.

The decisions about which companies to buy, hold or sell tend to be made by investment staff, or if outsourced, asset managers. Here lies another principal–agent relationship.

The asset manager has agency over the trustees. The asset manager must make investment decisions in the best interests of the trustees, and ultimately, the savers. The details of the investment strategy are set out in contract, often called a "mandate". Trustees often appoint investment consultants to provide support in selecting and monitoring the asset manager.

If an agent fails to make an investment decision in the best interests of the principal, the principal could take legal action, to be settled by the courts.

Fiduciary duties exist in common law countries, such as the UK or US. In civil law countries, such as France or Germany, there exists similar duties, although their application is less relevant to investment decision-making, with more prescription and regulatory intervention, rather than retrospective legal dispute determined by the courts.

In the relationships between asset manager and asset owner and between asset owner and saver, fiduciary duties tend to be interpreted as, "in the best financial interests". As such, if the activity is non-financial, then fiduciary duties preclude their consideration. Or in other words, for both the asset owner and asset manager, responsible investment must be proved to be in the financial interests of the saver.

Fiduciaries are sensitive to retrospective legal action where they have (or are seen to have) imposed their own personal views on their investment decisions.

In the case of a dispute—as a 2022 UK Department for Work and Pensions (DWP) guidance says—"Neither DWP nor TPR [The UK Pensions Regulator] can provide a definitive interpretation of the legislation which is a matter for the courts" (DWP 2022).

This interpretation is particularly acute in the US. Some aspects of responsible investment are considered political and not without reason. Climate change in the US is highly partisan. As is, but perhaps to a lesser extent, labour rights in the UK.

Case law is thin. Lawyers often cite Cowan vs. Scargill, an English trusts law case from 1984 (Sackers 1984). Arthur Scargill, then-president of the National Union of Mineworkers, sought to disinvest the National Coal Board's pension fund from industries competing with coal. The court found against it. The judge said that "the best interests of the beneficiaries are normally their financial interests".

A follow-up memorandum said, "Trustees … may be held liable for investing in assets which yield a poor return or for disinvesting in stock at inappropriate times for non-financial criteria" (Cowan vs. Scargill 1985).

As such, the interpretation of fiduciary duties was (and in some cases, remains) a threat to responsible investment. Here, the agent (typically, the trustees or the asset manager) interprets ESG issues, and perhaps sustainability more generally, as non-financial and therefore not in scope.

Increasingly, the argument that fiduciary duty precludes responsible investment has been turned on its head. Rather, fiduciary duties require responsible investment.

The UK Law Commission is independent to the UK government and has assessed responsible investment and fiduciary duties over many years. Its role is to "keep the law under review" and "recommend reform where it is needed" (Law Commissions Act 1965). In 2014, the Law Commission was tasked to evaluate "the extent to which fiduciaries may, or must, consider:

a. Factors relevant to long-term investment performance which might not have an immediate financial impact, including questions of sustainability or environmental and social impact;
b. Interests beyond the maximisation of financial return;
c. Generally prevailing ethical standards, and/or the ethical views of their beneficiaries, even where this may not be in the immediate financial interest of those beneficiaries" (Law Commission 2014).

The Commission concluded:

"Whilst it is clear that trustees may take into account environmental, social and governance factors in making investment decisions where they are financially material, we think the law goes further: trustees should take into account financially material factors."

"However, we do not think it is helpful to say that ESG or ethical factors must always be taken into account. These labels are ill-defined and liable to cause uncertainty."

"It is for trustees' discretion, acting on proper advice, to evaluate these risks" (Law Commission 2014).

The Commission proposed a two-stage test. First, "Do trustees have good reason to think that scheme members share the concern?," second, "the decision should not risk significant financial detriment."

The Law Commission's report was generally helpful for responsible investment. Three challenges however remained.

1. How would a trustee be expected to determine the views of scheme members? In the case of a pension scheme, there may be many savers. Some pension funds have millions of savers, representing a large proportion of the public, across gender, age, geography and level of wealth, and, presumably, a range of views on sustainability topics.

2. What is meant by significant in "significant financial detriment"? What if the detriment is in the short term but leads to value creation in the long term?
3. What guidance is there for a trustee in determining financial materiality? We tend to measure financial performance using historical data sets. But if trustees wait to review past performance then the decision of the extent to which the trustee should consider ESG issues would presumably be too late.

In the first chapter, we defined materiality. But given its importance to how we interpret fiduciary duties, we'll explore materiality in more detail, and in particular, whether we can prove that responsible investment contributes to superior risk-adjusted returns.

Financial Materiality

Nearly always, the primary argument in favour of responsible investment is that it contributes to superior risk-adjusted returns. In short, that a responsible investment strategy outperforms a non-responsible—or traditional—investment strategy.

Much of the academic research that attempts to address this question assesses ESG integration (as opposed to responsible investment), where the investor integrates ESG issues in investment decision-making in order to provide an investor with a deeper understanding of a company's financial prospects.

There are three, very different often-cited studies that provide useful, if dated, evidence.

The first is a meta-study published in December 2015 by Deutsche Asset and Wealth Management and the University of Hamburg (DWS 2015). The study investigates whether integrating ESG issues into the investment process has had a positive effect on corporate financial performance, whether the effect was stable over time, how a link between ESG issues and corporate financial performance differs across regions and asset classes and whether any specific subcategory of E, S or G had a dominant influence on corporate financial performance.

The results show that only 10% of the studies display a negative relationship with an overwhelming share of positive results, of which 47.9% of studies and 62.6% of meta-studies yield positive findings.

The second is a correlation study focusing on the US. In December 2017, MSCI and PRI presented three US studies. The three studies validated that ESG issues are materially linked to both equities and fixed income performance, with the degree of financial materiality varying across individual sustainability factors, fundamental profiles and industry groups (PRI 2017). The study also found that companies with "ESG momentum" tended to outperform.

The publication of this study followed the election of President Trump. The PRI was concerned that the US Department of Labor (DoL) would overturn rule-making introduced in the final months of the Obama administration that clarified that ESG issues should be integrated in pension fund decision-making. If ESG issues were financially material, the PRI reasoned, the rule-making was not necessary.

The Trump administration did overturn the rules; overturned once more by the Biden administration, which led many decision-makers (reasonably in my view) to interpret the rules as political, which inevitably favours status-quo bias. In other words, investment decision-makers concluded that the rules will keep changing and so there is no point in doing anything. There remain billions of dollars invested on behalf of US savers that do not consider ESG issues.

The third is a study of mutual funds, published in June 2019. Research by Morgan Stanley conducted on the performance of nearly 11,000 mutual funds from 2004 to 2018 showed that there is no financial trade-off in the returns of sustainable funds compared to traditional funds, and they demonstrated lower downside risk.

The study was refreshed in September 2020 (Morgan Stanley 2020). From January to June 2020:

- "U.S.-based sustainable equity funds outperformed their traditional peers by a median of 3.9%.
- U.S.-based sustainable taxable bond funds outperformed their traditional peers by a median of 2.3%."

When asked about financial materiality (and I often am), my stock answer is as follows:

"There are more studies than not that demonstrate that the integration of ESG issues contributes to superior risk-adjusted returns. Each responsible investment strategy and decision, however, should be assessed on its own merits."

I also refer to the World Economic Forum Global Risks Report (WEF, updated annually). While the backward-looking studies have some use, sustainability is inherently forward-looking.

The Global Risks Report does not prove financial materiality, however, year after year, sustainability issues top the list of issues, including climate action failure, extreme weather, biodiversity loss, livelihood crises, social cohesion erosion and more recently, cost of living.

When considering the evidence base on financial materiality, there is however a range of challenges. They include:

1. The inability to isolate ESG issues and to attribute financial performance. ESG issues are sometimes considered a proxy for quality. ESG issues represent a wide range of potential, and often, unconnected considerations—from board governance to climate risk. Correlation studies do not prove causation.
2. There are a range of approaches to responsible investment. The strategies may be badged responsible but the approach is conflicting.
3. There is little correlation between ESG scores across ESG data providers. As such, identifying high-performing ESG companies is not straightforward. Even if ESG data was aligned, financial analysis of ESG data differs. Even if the financial analysis was aligned, views on financial relevance differs. So it comes back to process.

There are of course scenarios where a sustainable company under-performs or a non-sustainable company over-performs. Inevitably, there is a flurry of gotcha-style articles.

My own view on the debate about whether ESG issues are financially material is that there are arguments for and against. For those arguing against, it is likely a proxy for more deeply held anti-ESG views.

Those with fiduciary duties, such as pension fund trustees, tend to set a high bar when it comes to the evidence base.

For readers looking for further research on ESG issues and financial materiality, I'd recommend Alex Edmans' research. Edmans is Professor of Finance at London Business School. His February 2023 paper, "Applying Economics – Not Gut Feel – To ESG" argues that approach to ESG issues should be grounded in economics, rather than instinct, driven by data and aligned with a company's business model.

The PRI also hosts a database of academic papers, many of which speak to financial materiality (PRI 2022).

I asked Alex Edmans to what extent can responsible investment help or hinder investment returns? "That itself is an interesting question because many people think of the link between ESG issues and company performance" Edmans said.

"But you specifically asked about investment returns and here there's an extra hurdle because for something to affect investment returns not only does it need to boost company performance, but also that the market doesn't currently take it into account."

"So one might argue that because of the attention to ESG it has actually made it harder to outperform."

I find this to be often misunderstood by responsible investment professionals. ESG issues may materially affect a company's performance, but that is often already priced by the market.

"So why might it be that ESG conceptually leads to higher investment returns?" Edmans asked.

"You need two things to be the case. 1) The ESG factor improves long-term financial performance of the company and 2) it also needs to improve performance of a company in a way that's not fully priced into the stock market."

"As such, intangible ESG issues are more likely to lead to higher returns, for example employee satisfaction, rather than tangible measures, like percentage of ethnic minorities on the board, because that's something which is really easy to measure, so it's quite likely that the market would have already captured that."

Take two companies:

- An unsustainable company where ESG hinders company performance, but the market prices the company as though the ESG issue has been addressed (the improver potential is priced in).
- A sustainable company where ESG helps company performance, but the market does not price the company as though the ESG issue has been addressed (the sustainability performance is not priced in).

Whether the investment strategy outperforms is subject to the skill of the investor's processes and ability to identify unpriced sustainability issues.

Nevertheless, ignoring sustainability altogether is clearly not in the best interests of the fund's performance.

Fiduciary Duty in the 21st Century

The PRI and UNEP FI's work on fiduciary duties, which started in late Spring 2014 at the Palais in Geneva, was accompanied by reams of legal analysis by US law firm, Latham and Watkins LLP. Latham and Watkins told us that they would not put their name to pro bono work, the rationale being that publicity could help (or hinder) commercial relationships, in which case, it would no longer be pro bono (but, marketing).

Latham and Watkins deserve credit. The contribution of their lawyers to the project was considerable.

In late 2014 and early 2015, we interviewed dozens of investors. The project team was Rory Sullivan, an external consultant, and now CEO and founder of Chronos Sustainability, UNEP FI's Elodie Feller and me.

We had a script, we asked questions and we prepared short summaries of our interviewees' comments. The investors represented a range of geographies and a range of types of investor—asset owner, asset manager, lawyer and service provider.

Some investors told us what we were hoping to hear. Some did not. But when we came to conclude our assessment, it was clear we weren't too far off the mark. Most of the investors we spoke to agreed that investment requires the integration of ESG issues.

We also went through a peer review process. The feedback from the World Bank was negative. The report, they said, was not well-written, and seemingly, unhelpful. We organised a follow-up call and their feedback helped strengthen our analysis. A few years later a PRI and World Bank partnership on responsible investment policy ensued.

It was the feedback from State Street's legal counsel that was most welcome. State Street said the report was "directionally correct".

As authors we reported to a steering group including PRI's and UNEP FI's CEOs. We met in PRI's boardroom to agree the report's findings. For all the evidence we'd collated, we still debated whether ESG integration was a can, should or must. On our recommendation, the steering group went for must.

The evidence we'd collated allowed us to conclude that investors must integrate ESG issues as a requirement of their fiduciary duties. This is because ESG issues contribute to the investment thesis. Integrating ESG issues allows an investor to optimise risk-adjusted returns.

Over the weekend, I sat down to write the report's executive summary.

The purpose of the report, we argued, was to end the debate about whether fiduciary duty is a legitimate barrier to investors integrating ESG issues into investment processes. I spent some time on the line, "failing to integrate ESG issues in investment practice is a failure of fiduciary duty." It was the hardest hitting interpretation of the word "must" that the steering group had agreed to.

The report set out what fiduciary duties are, their prevailing interpretation and the relevant investment practice and regulation across eight countries that led to our conclusion.

Its precursor, a 2005 report commissioned by UNEP FI from law firm Freshfields concluded that integrating ESG considerations into investment analysis is "clearly permissible and is arguably required" (UNEP FI 2005). We wanted Fiduciary Duty in the 21st Century to go further.

"Far from being a barrier," we said, "the report finds that there are positive duties on investors to integrate ESG issues, to mitigate risk and identify investment opportunities."

One of our interviewees, Paul Watchman, Honorary Professor at Glasgow University's School of Law and author of the 2005 Freshfields report said, "The concept of fiduciary duty is organic, not static. It will continue to evolve as society changes, not least in response to the urgent need for us to move towards an environmentally, economically and socially sustainable financial system" (PRI 2015a).

Despite significant progress, we found that outdated perceptions about fiduciary duty persisted. Many large investors had yet to fully integrate ESG issues into their investment decision-making processes. Lawyers and consultants, particularly in the US, too often characterised ESG issues as non-financial. This is still the case. We also found that there was inconsistency in corporate reporting and weaknesses in the implementation of legislation and codes on responsible investment.

To modernise interpretations of fiduciary duty in a way that ensures these duties are relevant to twenty-first-century investors, the report proposed a series of recommendations for institutional investors, financial intermediaries and policymakers.

Institutional investors, we said, should make explicit their commitments to ESG integration, implement across investment processes and require companies to provide ESG reporting. Intermediaries should advise their clients to take account of ESG issues in their investment processes. Policymakers should harmonise definitions of fiduciary duty and clarify that fiduciary duty requires that investors pay attention to long-term value drivers.

These relatively modest changes in interpretation and practice of fiduciary duty could catalyse rapid change in the importance assigned by investors to ESG issues.

Contributing to the report with a foreword, Al Gore and David Blood concluded, "there is no 'do nothing' task … considering sustainability is not only important to upholding fiduciary duty, it is obligatory. Sustainability is an important factor in the long-term success of a business."

The report was a success. We launched at PRI's 2015 conference. What followed was a multi-year, multi-million pound research programme to upgrade pension fund regulation where ESG integration was optional to ESG integration was required, over-turning the narrative that fiduciary duties preclude ESG integration.

We convened a global investor statement supporting our findings, with well over 100 signatories. We organised investor and policymaker workshops on our report in every major financial centre. We developed detailed policy recommendations and took our findings to policymakers.

It's fair to say, it's a task that's still underway.

PRI In Person 2015

The 2015 PRI annual conference took place in the UK, in London's Excel (unfortunately for PRI, alongside an arms fair).

The conference was attended by 1000 people; this in itself was a milestone. That 1000 people were sufficiently involved in responsible investment to attend a three-day, fee-paying conference marked the end of the beginning for responsible investment. Responsible investment had come of age and it wasn't going away.

The PRI had recently hired a new head of press. The PRI's major publication for the conference was Fiduciary Duty in the 21st Century.

For the first time, a PRI report launch was accompanied by a media strategy and widely covered in the trade press.

Shortly after its launch, an email was sent by a head of responsible investment at a UK investor (I'm not going to name him) to another 20 or so similarly titled people at other investors with a link to the report that we'd just launched. The accompanying text said, and I quote, "what the fuck?" The sender disagreed with the report's findings. It wasn't, of course, sent to me, but it was forwarded.

Today, most investors will accept that some form of ESG integration is mandatory, but it wasn't always the case. Indeed, some responsible investment professionals thought the conclusions undermined their positions. It moved responsible investment from a standalone team to the investment team. Rightly so.

I met the sender of the email for a coffee a few weeks later. True to our Britishness, he didn't mention the email and nor did I. Rather we had a pleasant coffee and discussed a range of topics, somehow managing to agree on what we thought was meant by responsible investment and fiduciary duties.

In other words, it was bark with no bite. The findings, which we thought may have been controversial, were largely accepted.

The report was formally launched at a panel discussion I moderated. The director of The Generation Foundation attended.

The next day, we took a hastily prepared programme of work to The Generation Foundation's Mayfair offices. With appropriate funding we said we would take the country case studies in Fiduciary Duty in the 21st Century as our starting point and develop roadmaps with policy recommendations, which we'll work with policymakers to implement.

To our surprise, Generation agreed, and we got to work.

References

Cowan vs. Scargill (1985), Ch 270.

DWP (2022), Reporting on Stewardship and Other Topics Through the Statement of Investment Principles and the Implementation Statement: Statutory and Non-Statutory Guidance. [online]. Available from: https://www.gov.uk/government/consultations/climate-and-investment-reporting-setting-expectations-and-empowering-savers/outcome/reporting-on-stewardship-and-other-topics-through-the-statement-of-investment-principles-and-the-implementation-statement-statutory-and-non-statutory.

DWS (2015), ESG & Corporate Financial Performance: Mapping the Global Landscape. [online]. Available from: https://download.dws.com/download?elib-assetguid=2c2023f453ef4284be4430003b0fbeee (Accessed, January 2023).

Edmans (2023), Applying Economics—Not Gut Feel—To ESG. [online]. Available from: https://papers.ssrn.com/sol3/papers.cfm?abstract_id=4346646 (Accessed, February 2023).

Law Commissions Act (1965). [online]. Available from: https://www.lawcom.gov.uk/ (Accessed, January 2023).

Law Commission (2014), Fiduciary Duties of Investment Intermediaries. [online]. Available from: https://www.lawcom.gov.uk/project/fiduciary-duties-of-investment-intermediaries/.

Morgan Stanley (updated 2020), Sustainable Reality Analyzing Risk and Returns of Sustainable Funds. [online]. Available from: https://www.morganstanley.com/content/dam/msdotcom/ideas/sustainable-investing-offers-financial-perfor mance-lowered-risk/Sustainable_Reality_Analyzing_Risk_and_Returns_of_Sust ainable_Funds.pdf (Accessed, January 2023).

PRI (2015a), Fiduciary Duty in the 21st Century. [online]. Available from: https:// www.fiduciaryduty21.org/publications.html (Accessed, January 2023).

PRI (2017), Financial Performance of ESG Integration in US Investing. [online]. Available from: www.unpri.org/download?ac=4218 (Accessed, January 2023).

PRI (2022), Academic Directory. [online]. Available from: https://www.unpri.org/ academic-directory/10653.article (Accessed, February 2023).

Sackers (1984), Cowan vs. Scargill (High Court)—13 April 1984. [online]. Available from: https://www.sackers.com/pension/cowan-v-scargill-high-court-4-april-1984 (Accessed, January 2023).

UNEP FI (2005), A Legal Framework for the Integration of Environmental, Social and Governance Issues into Institutional Investment. [online]. Available from: https://www.unepfi.org/fileadmin/documents/freshfields_legal_resp_2 0051123.pdf (Accessed, January 2023).

WEF (updated annually), Global Risks Report. [online]. Available from: https:// www3.weforum.org/docs/WEF_The_Global_Risks_Report_2022.pdf (Accessed, January 2023).

5

Less Why, More How

Growth of ESG Analysts

From 2015 to about 2018 responsible investment grew.

Investors set up responsible investment teams, hired ESG analysts, took out contracts with ESG data providers and started to market their sustainability credentials.

Seemingly overnight, responsible investment became commercial. It allowed investors to differentiate their offerings.

Quick off the mark was MSCI. MSCI is a US-based index provider, research provider and more recently, ESG and climate solutions provider. The MSCI "credit card" could often be found behind the bar. I'm sure MSCI would attribute their growth to their products. But MSCI was one of the first major service providers to identify sustainability as a revenue opportunity, and therefore, prepared to sponsor research (and drinks). It was quite the break from the "tea and biscuits" events of old.

Responsible investment was in vogue. Working in responsible investment evolved. It was less "why" and more "how".

In responsible investment's early days, ESG analysts were given free rein. Responsible investment was undefined, potentially even vague. To an extent, ESG analysts could make their role what they wanted it to be.

But from around 2015 onwards, ESG issues became more important to investment processes, with ESG analysts increasingly contributing to investment decision-making, providing input and ESG data to portfolio managers.

© The Author(s), under exclusive license to Springer Nature
Switzerland AG 2023
W. Martindale, *Responsible Investment*, https://doi.org/10.1007/978-3-031-44536-1_5

ESG analysts tended to be one of two types. For the more traditional firm, it was often a "mainstream" analyst, with an interest in sustainability. This was more typical in the US. Asset managers would tolerate this, often as little more than a hygiene factor, perhaps as a nod to their European clients. The analyst tended to be well-paid, although, anecdotally at least, bonuses for ESG analysts tended to lag their non-ESG counterparts.

The other route was via an NGO. This was more typical in Europe and the UK. NGOs are less partisan in Europe than in the US. The pay for a responsible investment analyst was better than an NGO, but very much on the low side within the financial sector. This type of ESG analyst tended to have more sustainability-related expertise but be on the periphery of investment decision-making—a standalone person or team, outside day-to-day investment decision-making.

Since around 2018, this started to change (I'm generalising here, there were and are exceptions both to the type of ESG professional and the timing, but this is the trend I was aware of at the time). ESG analysts became "heads of responsible investment" or even "Chief Responsible Investment Officers", their reporting line moving from operations to investment, and increasingly, directly to CIO or even CEO.

There were a range of ESG-related roles emerging, which persist to this day.

1. The company analyst. This is a traditional analyst role, through the lens of sustainability, assessing a company's sustainability footprint, and the way in which sustainability is managed, to identify financially material sustainability-related risks or opportunities that are not well-reflected in the company's price. For data providers and active managers this is an important role.
2. The thematic analyst, often in the form of jurisdictional, sector or macro-economic experts. This role identifies sustainability-related themes that will affect the financial performance of companies exposed to that theme. For example, an energy analyst, an expert on biodiversity or an expert on human rights.
3. Engagement and voting analysts. Diversified investment strategies can include more than 500 companies, each with an AGM, director nominations and increasingly, resolutions on sustainability topics. Voting is in part administrative, preparing an engagement policy and a voting policy, ensuring voting is consistent with the policy, but it is also high impact. Engagement meetings, filing resolutions and voting against directors have

materially changed company behaviour. When done well, this is a high impact role.

4. The specialised impact expert. This is by far the "coolest" of the sustainability-related roles, often flying around the world to assess potential investments that provide material real-world benefits—fishing communities in Indonesia, forestry in Kenya or solar technology in California.

5. Finally, reporting. Most investors are subject to reams of reporting requirements—PRI, NZAMI, stewardship code, TCFD. The list goes on.

Often, such is the nature of the sustainability industry, the role is a mix.

In my experience, sustainability professionals tend to work long hours. I think this is often the case for vocational roles. The world is unsustainable and appears to be getting more unsustainable. We feel responsible, because we have agency. We can make a difference, so we better work that bit harder, otherwise climate breakdown is on us. At least, that's what it can feel like.

I've developed a good network in responsible investment, because I find I can more quickly get to grips with a topic if I can speak to someone with expertise.

This involves an "ESG dinner club", several WhatsApp chats, attending conferences or webinars, as well as participating in PRI and IIGCC working groups. I enjoy this, but it does put pressure on other parts of my responsibilities.

This means that, sometimes the "day job" is far from a day job, and gets taken care of in the evenings or at weekends. I think this is typical for those in similar roles.

Head of Sustainability

In January 2023, while I was working at UK and Dutch investor, Cardano, I was asked by the UK's Financial Conduct Authority (FCA) to put together a few ideas on "how a chief sustainability officer can most effectively support a firm in achieving its climate- and sustainability-related objectives."

It was the first time I was asked to formalise my views on how to be successful in my role. The final text differed a little, subject to the FCA's editorial teams. But here's the original:

1. Lead from the top: It's important to establish CEO leadership on sustainability topics. The head of sustainability should report to the CEO or

CIO. This provides the head of sustainability with the mandate necessary for fast-track implementation.

2. Here are some of the potential workstreams that could contribute to a project plan.

- Beliefs and policy
- Sustainability data
- Regulation
- Investment
- Exclusions
- Engagement and voting
- Stakeholder groups
- Education
- Client training
- Reporting.

For the head of sustainability, it is important to provide expert input—but not necessarily "to do". To be successful, it's important to embed sustainability across your company. That means that all teams are responsible for sustainability, even if "sustainability" isn't in their team name or job title.

3. Ensure business-wide involvement.

To ensure sustainability is encompassing, consider establishing "sustainability champions" in every team (someone in the team that leads on sustainability topics). This should be voluntary, and I expect there will be more candidates than roles, reflecting employee interest in sustainability.

Support sustainability champions by organising training sessions on a range of topics. Consider external speakers, such as from PRI, IIGCC or human rights NGOs. Maintain a sustainability reading list, as well as suggested podcasts and films.

Examples include:

- How to avoid a climate disaster, by Bill Gates (2021).
- Just Business, by John Ruggie (2013).
- Doughnut economics, by Kate Raworth (2017).
- Impact, by Ronald Cohen (2020).
- The truth about modern slavery, by Emily Kenway (2021).
- Making the Financial System Sustainable, edited by Paul Fisher (2020).
- The Social Licence for Financial Markets, by David Rouch (2020).

Post-pandemic, most companies continue to organise "all staff calls". Ensure sustainability is a frequent agenda item to take your colleagues with you.

4. Establish beliefs and policies.

Establish sustainable investment beliefs ("the why") and sustainable investment policies ("the how").

Review the evidence base on financial materiality and prepare client training materials to support your clients in understanding your approach to sustainability.

Next review terminology and approach, in particular, what is your approach to "real-world sustainability impact" or "double materiality".

To support the development of beliefs, you may find it useful to organise internal workshops, as well as discussions with clients, to explain what you mean by sustainability, the rationale for your approach and the implications for your investment decisions.

To define what is meant by sustainability, you could use the definitions of sustainability and stewardship taken from the United Nations Brundtland Commission (1987), the Sustainable Development Goals (2015) and the UK Financial Reporting Council (FRC) stewardship code (revised 2020).

Finally, set out your position on decarbonisation and net zero.

5. Support portfolio managers with data and metrics.

ESG data can be overwhelming. ESG data itself is not useful unless it is understood by investment decision-makers and integrated into investment and client-reporting processes.

To identify the metrics, assess a range of industry standards, including:

- PRI's Inevitable Policy Response to inform your climate change scenarios.
- Institutional Investor Group on Climate Change (IIGCC) and Paris Aligned Investment Initiative (PAII) to inform your GHG emissions metrics.
- And a combination of PRI programmes, ShareAction, Partnership for Carbon Accounting Financials (PCAF) and Partnership for Biodiversity Accounting Financials (PBAF), as well of course, your own areas of interest to inform your ESG metrics.

It isn't always necessary to subscribe to third-party data providers. The latest AI technologies can help here. It's important that investment teams undertake their own GHG and ESG analysis of investments, prior to decision-making and on an ongoing basis.

For the voluntary frameworks, consider what is right for you, and what is not. For the regulatory frameworks, undertake a gap analysis, what you're currently doing and what you need to do.

Where possible, align with industry groups.

6. Set out your approach to real-world sustainability impact

Investors increasingly incorporate both ESG integration and real-world sustainability impact into their investment processes. However, there is little clarity about what is meant by influence, real-world sustainability impact, and how to measure it.

7. Where there is expertise, step aside.

There will be more work to do than people to do it. So where there is already expertise, step aside.

8. Report.

Report to regulators, clients, stakeholder groups—and internally. Reporting, for example, via your website, allows you to take your colleagues and clients with you. Undoubtedly, there will be some scepticism, or even opposition, and transparency enables you to set out your rationale for the decisions you're making.

9. Innovate and change.

In my experience, the role of head of sustainability is constantly evolving, with new sustainability themes, new collaborative engagement initiatives, new disclosure obligations and new client expectations.

10. Invest in relationships.

Sustainability can be exclusive. Terminology changes. Expectations evolve. There are new groups, new requirements and new reporting. These are barriers that need to be addressed—and it is the role of the head of sustainability to do so.

When I originally published these 10 recommendations it was the following paragraph that received the most pushback:

When meeting with graduates or students, or more senior investment professionals considering a change in career, I'm often asked for my views on what to study or read: "What about the CFA sustainability qualifications?" "What about the Oxford or Cambridge sustainability course?" Studies are of course important. But in my view, the most important characteristic is passion. To be successful as a head of sustainability, the starting point has to be a passion to change markets, economies and societies to be more sustainable—to make the world a better place—and a belief that investment is a key conduit to achieve that change.

I asked Roger Urwin whether he agreed. He did, but only up to a point. "One of the key dimensions is about the passion through which people pursue sustainability. That's both its most important element, and its biggest weakness. It can distract from the pureness of thinking that's needed. It becomes green-wishing if you take it too far."

ESG Integration

While a modern interpretation of responsible investment goes beyond ESG integration, from 2015–2018, ESG integration was foundational. I think that the best articulation of ESG integration (to this day) was set out in a Fiduciary Duty in the 21st Century US Roadmap published in September 2016, a few weeks before Trump's election victory (PRI 2016).

It says:

"The PRI defines ESG integration as the systematic and explicit inclusion of material ESG factors into investment analysis and investment decisions."

"… investors can treat ESG factors in the same way as any other financial factors using existing quantitative methodologies."

"Neglecting analysis of ESG factors may cause the mispricing of risk and poor asset allocation decisions. It is worth clarifying that ESG integration does not necessarily involve a narrowing of the available investment universe (unlike negative screening). Neither does it involve relegating the pursuit of a financial return to unrelated objectives (social or ethical)."

"It does though provide investors with an expanded set of tools for evaluating the operational performance and financial prospects of investee companies. ESG analysis is increasingly assisting investors to identify value-relevant factors undetected by outdated financial-only analysis."

"… It is important to understand that ESG integration is less a product than it is part of the broader process and technology of investment analysis. Ultimately, the consideration of ESG factors has become one of the core characteristics of a prudent investment process."

"Investors also face a number of systemic risks, such as the … impacts of climate change. Climate change may significantly alter the investment rationale for particular sectors, industries and geographies and may have generalized negative impacts on economic output."

"… Systemic risks, particularly those arising from ESG issues, should be pro-actively identified and assessed as part of prudent investment decision-making."

"Investors should adjust their analytical capacities to identify, integrate and provide transparency on their management of such long-term systemic risks."

There are other explanations, but this is the most succinct, and covers the key points. ESG integration is not a product. Rather ESG integration is part of the process of investment decision-making. It is not ethical, but financial. It includes systemic risks, such as climate change. ESG integration is about financial performance.

ESG integration might have real-world impact. Investors might decide to invest in a company that does address ESG risks and not to invest in a company that does not.

All things being equal, this will mean more buyers of companies that address ESG risks, and less buyers of companies that do not. This will push up the value of the company that does and lower the value of the company that does not address ESG risks.

In theory at least, this will affect the cost of capital. In this scenario, companies that address ESG risks will be able to borrow at preferential rates and expand their business, at the expense of the company that does not address ESG risks.

The results of studies that attempt to determine whether this happens in practice are mixed. ESG integration may lead to real-world impact, but almost certainly other conditions must be in place, including regulatory change and consumer pressure.

Given recent focus on ESG issues, some investors consider ESG issues over-valued, hence the term, "the ESG bubble". They say investors have over-prioritised ESG issues at the expense of other financially material issues.

As with all market transactions, the price will reach an equilibrium matching buyers and sellers. Some buyers may be less likely to buy if they consider a company has unpriced ESG risks, but other buyers will step in, seeing the fall in price, and consider the ESG risks adequately priced.

Some investors go further saying that by selling the company the investor loses influence.

ESG integration is not a binary construct, but a scale, where investors, to differing degrees, consider ESG issues relevant to a company in investment decision-making.

Stewardship

Accompanying ESG integration is stewardship. It is no coincidence, that in PRI's six Principles, ESG integration is first, stewardship second. Principle 2 says, "We will be active owners and incorporate ESG issues into our ownership policies and practices."

Typically, investment teams are structured accordingly, with resourcing on ESG integration and stewardship, although often biased in favour of ESG integration.

EU regulation is equally biased, with substantial time and effort spent on defining sustainable funds and fund-level disclosures, all subject to political negotiation across Council, Parliament and Commission. Meanwhile the Shareholder Rights Directive (SRD) receives substantially less attention.

In the UK there is more attention to stewardship for two main reasons.

1. The UK's legal system is common law, with investment decision-making subject to fiduciary duties. Rather than exclude unsustainable companies, investors tended to invest (for fear of breaching fiduciary duties), but use stewardship to achieve sustainability objectives. In recent years, the difference in approach between the UK and EU has blurred.
2. In 2011, following the global financial crisis, the UK government commissioned Professor John Kay to review the functioning of equity markets. The first of Kay's recommendations was: "The Stewardship Code should be developed to incorporate a more expansive form of stewardship, focussing on strategic issues as well as questions of corporate governance" (Kay 2012). The report says, "Promoting good governance and stewardship is therefore a central, rather than an incidental, function of UK equity markets." "Regulation should favour investing over trading." The UK Stewardship Code has since undergone several reviews.

Stewardship, also known as active ownership, typically comprises engagement and voting. Engagement can take place all year round and takes many forms.

In the past, stewardship was mostly undertaken by investors with controlling, or perhaps large minority, stakes. Over the past 25 years, stewardship has improved considerably, although the prevailing approach to stewardship on ESG issues is a long way from reaching its potential.

Engagement ranges from emails to calls to meetings. It can include stakeholder engagement (such as NGOs or industry groups), media engagement or, as a means to escalate, voting, filing resolutions or even legal action. A typical engagement meeting is between staff at a company's investor relations team and staff at an investor's stewardship team, but there are variations of this.

Portfolio managers (those directly responsible for investment decisions) will often speak with companies to better understand company decision-making, and the line between portfolio manager-led engagement and stewardship team-led engagement is often blurred. If I had to generalise, stewardship professionals tend to engage on sustainability themes, portfolio managers on governance themes or for informational purposes.

Engagement can also be at more senior levels, between investor CEO, CIO or board and company CEO, CFO or board. Meetings tend to be private. Many investors will provide statistics on meetings (how many engagement meetings and how many companies) but rarely provide details of what was discussed.

Not all engagements are equal. Some investors employ experienced, senior stewardship professionals, with clear stewardship objectives, escalation strategies, interim milestones and reporting to clients and stakeholders, where the stewardship objective is an outcome—a change undertaken by the company. Some do not.

Whereas engagement tends to be private, voting is public, so assessing stewardship can often focus on voting decisions. Shareholder meetings take place once a year, for much of Europe and America that's around March, April and May.

Shareholder meetings are required by law and include votes on the company's financial statements, accounts and the election of directors, and sometimes, topics such as the company's decarbonisation strategy or the publication of a company's TCFD report.

A vote is binary and therefore misses the nuance of engagement. For example, a vote against a company's decarbonisation strategy may be because the investor considers the strategy too ambitious, or it may be because the investor considers the strategy not ambitious enough.

Investors tend to rely on proxy advisors, a company that provides research, advice and voting recommendations to investors (for a fee). Proxy advisors will have their house view, and then provide a series of options on ESG topics to suit the investor.

More sophisticated investors will provide their proxy advisor with their own voting policy, and require the proxy advisor to make recommendations accordingly, assessing the proxy advisor's recommendations, and if necessary, over-ruling.

Shareholders are entitled to file, or co-file, at shareholder meetings. This involves writing, submitting and seeking support for a resolution, for example, calling on the company to pay its staff a living wage.

Stewardship codes are evolving to support and encourage good-practice stewardship. The first iteration of the UK Stewardship Code, published before Kay's report in 2010, was revised in 2012, and then again in late 2019. The Code is voluntary, although it is referenced in FCA regulation, and most UK large equity investors and major overseas investors are signatories.

The Code is regulated by the Financial Reporting Council (FRC), which oversees corporate governance, as well as stewardship.

The FRC has made efforts to improve disclosure against the Code. It first introduced tiering, signatories to the Code were tiered depending on the quality of their submission, but investors were quick to understand the tiering framework, and most would achieve the top tier.

Following the revision to the Stewardship Code in 2019, around a third of applicants, which were submitted in March 2021, failed to meet the FRC's standards, including Schroders and JPMorgan Asset Management (FT Adviser 2021).

This is mostly due to reporting, rather than process, but nevertheless, the FRC's review process is impressive. For the 2021 submission, the FRC reviewed a draft submission, highlighting weaknesses and strengths using track changes and comments. The FRC then provided bespoke feedback on the final submission, setting out-performance expectations for the following year's submission.

The results are published privately a few days prior to their public release. It makes for a nervous day, refreshing emails and waiting for the results. A few asset managers rejigged their staff in the weeks following the Stewardship Code results.

The FRC defines stewardship as follows:

> Stewardship is the responsible allocation, management and oversight of capital to create long-term value for clients and beneficiaries leading to sustainable benefits for the economy, the environment and society. (FRC 2019a, b)

The definition is applicable to asset owners (in their stewardship of asset managers) as it is to asset managers (in their stewardship of companies).

It is the final part of the sentence that is most noteworthy: "Stewardship should lead to sustainable benefits to the economy, environment and society." In other words, the purpose of stewardship is both to engage companies on their unmanaged financial risks, engage companies on risks systemic to portfolios and engage companies in pursuit of real-world impact to environment and society.

Those responsible for writing the code tell me that the text "to create sustainable value for beneficiaries, the economy and society" drew the most attention, both for and against, in the FRC's consultations. "Environment" wasn't in the first iteration, but was in the final version, supported by consultation responses (FRC 2019a).

The FRC's feedback statement on the consultation (FRC 2019b) said, "Nearly all respondents also commented on the proposed definition of stewardship in response to this question. Approximately half of respondents commented that the primary purpose of stewardship is to deliver financial returns for clients. They acknowledged that in doing so there may be positive impacts for the economy and society, but that they did not see creating sustainable value for the economy and society as the primary aim of investor stewardship."

"By contrast, one third of respondents said that having regard to the economy and society in investment decision-making is necessary to properly fulfil their fiduciary duty. Some respondents called for 'the environment' to be included in the definition."

While, to some extent, stewardship has been a feature of equity markets for decades, stewardship is increasingly undertaken by credit investors too. Credit investors are not entitled to voting rights, but companies are incentivised to take meetings with their credit investors.

Bonds are traded on the secondary market. Pricing in the secondary market may affect whether the company can refinance at a lower rate of interest. Companies will be financing and refinancing on a regular basis.

Some investors subscribe to the notion, "engage equity, deny debt." In other words, withhold the debt of unsustainable companies. If enough investors act accordingly, it pushes up the price of issuing debt, incentivising

the company to take steps to be more sustainable—while, retaining investments in equity and using stewardship and voting to force the company to
take steps to be more sustainable.

Collaboration has long been a feature of stewardship. In the UK and EU,
policymakers have sought to clarify that conversations between investors on
sustainability topics do not constitute acting in concert. Persons "acting in
concert" are those with some form of agreement or understanding to coop-
erate to seek control of a company. However, this isn't the case in the US,
where there remains concerns that collaborative engagement does constitute
acting in concert. This interpretation has no doubt limited the extent of
investor collaboration on sustainability issues in the US. In research for this
book I was told that SEC insiders do not agree and consider this an excuse
put forward by US asset managers.

The PRI hosts a collaboration platform. It's a technology platform with a
log in that allows investors to propose a collaborative engagement or if it's
already underway, join a collaborative engagement. It's a neat idea, but with
few exceptions, it doesn't work well. Investors rarely collaborate through PRI
of their own accord. Collaborations tend to be most effective when coordinated by PRI staff or when investors already know each other well (and
therefore, do not rely on the collaboration platform).

Other groups, such as the UK Investor Forum do likewise. The Investor
Forum charge a fee, but it is a cost effective way for an investor to undertake
engagement.

My own experience of stewardship is not a particularly positive one. I've
participated in well-run thoughtful engagement meetings, where investors are
well-briefed, companies are engaged and the decision-making constructive.

But most meetings I've attended are not like that. Companies are not
incentivised to put forward senior staff. Investors on the whole are not well-
briefed. Investor escalation, such as disinvestment or votes against directors,
is unlikely.

And an issue that is largely unexplored by responsible investment stakeholder groups is that we (as people) tend to avoid conflict. There is a careful
balance in engagement between relationship-building and escalation. That
has become harder as meetings have shifted to be remote. Particularly so when
the meeting participants themselves are not the decision-makers.

Some engagements may be informational, but often, the engagement is
trying to get the company to do something it currently isn't. The company
may have good reasons why. But this is likely to be a difference of view, and
therefore, lead to disagreement.

Escalation strategies, supported at a senior level that do not constitute some form of conflict, are rare, if impossible.

Which is why, while investors go through the motions on stewardship, much of current practice is ineffective. The evolution of stewardship is still very much underway.

I asked Claudia Chapman, at the time Head of Stewardship at the UK's FRC, to explain the characteristics of good-practice stewardship.

"Being purposeful and intentional. Linking stewardship to your purpose and investment beliefs and aiming to be predominantly proactive in your approach."

"I agree with your view," Chapman told me, "that we can all be so well meaning in a very noisy space where it is difficult to have impact that we don't have the courage to say - you know what, X is doing a great job of Y. I am going to focus on Z. I will put my AUM behind Y, but I'll achieve more focusing on this one thing over here."

"Sometimes smaller firms seem to do it well. They are more nimble, more integrated and better able to articulate their purpose and investment beliefs, have a client base aligned to that and then stick by that identity in their investing."

And finally, "Working together. Coordinating, connecting and collaborating. Surely it makes sense to do this. One of the best things about the responsible investment profession is working with clever, dedicated and hardworking people in other organisations in this space. Irrespective of this, bringing together AUM to influence, or different resources, skills and experience from investors, especially if they are not large, should have a better outcome than each investor plugging away on their own."

Engage or Divest

Even today, conference organisers are setting up panels titled engage or divest. A tired "gotcha"-style question that is designed to divide the responsible investment industry, that was subject to much debate as responsible investment developed.

If the company is considered unsustainable, some asset managers engage, some asset managers divest. This sounds opposing, but in practice, most do both.

Those that primarily engage, will indeed divest if a company drops in value (which, may well be its fate, if it remains unsustainable and is on the wrong side of government regulation and consumer choice). Those that divest will often first engage.

In some circles, the question can prompt an earnest "we believe in engagement over divestment", the argument being that if you divest you "lose influence". This holds true to an extent, however, it assumes that all companies respond to engagement. Many do not. Divestment is a natural escalation.

In the years ahead, I expect investors will become more comfortable with divestment.

And companies that are open to engagement will likely meet with investors pre-investment and even post-investment as they would during investment.

Many investors will be prepared to "re-invest" if the issue that prompted the divestment is addressed. We are also seeing escalation after divestment, including engagement using media or even legal action, and so there remain forms of influence across the investment decision-making processes.

It may sound counterintuitive to engage having divested, but a minority of investors do, and they do it for two reasons.

1. To seek change such that the company becomes investable, wanting to grow the investable universe.
2. Even if divested, some companies can cause negative financial outcomes for other companies in the portfolio. An energy company that does not transition will supply high carbon energy to other companies in the portfolio.

Indeed, investors will likely invest across multiple strategies, some active, some passive, some equity, some credit, some public, some private, some physical and some derivative. It might be that they divest a company from some of their strategies, but not all of their strategies.

And some investors, particularly those with active investment strategies, will not have invested in the first place in order to divest.

In other words, engagement, investment, voting and divestment are all part of the toolkit that investors use as part of their integration of ESG issues in investment decision-making, and different investors will go about it in different ways, depending on their investment strategy, their approach to risk and their conviction on ESG issues.

Perhaps the real question is whether it's "divest" or "disinvest." Divestment means to sell in its entirety. Disinvestment means to sell part of one's holdings. But the words are used interchangeably.

The ESG Data and Ratings Agencies

Assessing a company's exposure to and management of ESG risks and opportunities is not straightforward for a few reasons:

1. Most investors will invest in hundreds, if not thousands, of companies.
2. Fundamental analysis is expensive, requiring analysts to review company reports, policies and practices. Depending on the depth of the analysis, one analyst will only be able to cover a few dozen companies.
3. Companies will disclose in a different way depending on how they are regulated.
4. Companies themselves are rarely comparable, operating at different depths through the intermediation chain. Some auto companies will, for example, manufacture tyres themselves, others will buy from a tyre company. The reported GHG emissions of the former will be higher than the latter. But, if you don't know the company, you won't know why.
5. Finally (and there are many more complexities), overseas companies will disclose in their local language.

For even mid-sized investors, this is an issue. It's simply not affordable to hire analysts who can assess such a volume of companies. Cue the ESG data and ratings agencies.

There are now dozens of ESG data and ratings providers. The largest are MSCI, Sustainalytics, S&P Global, Bloomberg, FTSE Russell, ISS, RepRisk, Clarity AI and Refinitiv. There are a number of more bespoke impact data providers, such as Impact Cubed or biodiversity specialist, NatureAlpha.

The data providers undertake a range of activities, condensing company disclosures, company press releases, company policies, public accounts, media articles, NGO reports, social media campaigns and much more into comparable, manageable indicators, such as GHG emissions, board diversity, incidents of human rights breaches and so forth.

The data tends to be in two forms.

1. Objective or reported data. Here, the data provider is simply reporting fact. There is no interpretation.
2. Subjective or interpreted data. Here, the data provider is interpreting fact into an indicator that the investor can use in their investment processes, such as a corporate governance score.

There is no universally agreed standard for corporate governance, so the data provider will assume their own standard, and then test companies against it, marking up or down based on the data sets obtained by the data provider.

Some, indeed, most companies' disclosures will be incomplete, so data providers may estimate data or use proxies. For example, say a Chinese cement manufacturer does not disclose its water use. The data provider may compare to other cement companies, based on the size and characteristics of the region in which it operates, as well as tons of cement manufactured.

ESG data and ratings agencies will compete based on the quantity of data, the quality of data and the availability of data necessary for regulatory disclosures. And of course, fee. The ESG data and ratings agencies are not cheap. A mid-sized asset manager is likely to spend as much as a million dollars in fees on data. The larger asset managers will spend much more.

Some data providers will employ teams of analysts, assessing company disclosures. More recently, data providers have turned to AI technology. Software will, what's called, "text mine" to identify relevant disclosures, and turn multi-page documents into decision-useful indicators.

Care here needs to be taken for accuracy, particularly for multi-jurisdictional, multi-language companies. Say a company's water policy covers its European but not its African supply chains, but it's Africa where water management matters most. While it's continually improving, some of the software I've seen is not clever enough to identify this.

In my experience, data providers provide substantially more data than investors use. Indeed, while many investors will say they would like more data, much of the existing data is unused. There's also the question of what to do with the data.

Most investors would use the ESG data providers as a starting point, supplementing the metrics with their own analysis. Most investors would also use more than one ESG data provider, combining the analysis or integrating what they consider the strengths of one provider, with that of another.

For "rules-based" passive investors, the data may inform index construction. For active investors, the data may be a starting point for fundamental analysis that informs security selection. The data is also used for reporting purposes, in particular, TCFD reporting in the UK or SFDR reporting in the EU.

Most data providers also provide a singular ESG score: A single, holistic assessment of a company's approach to sustainability.

I don't find the concept of a single ESG score compelling. Why add, say, methane emissions to labour rights to board diversity? The issues are unconnected. And each part of the score is itself subjective. Who's to say how much methane is acceptable per unit of production? And the weighting of each part of the score is subjective.

Nevertheless, investors like their metrics, and the ESG scores are simple to use and find their way into many ESG-badged investment products. For example, a feature of many ESG investment products is that the portfolio has an "ESG score higher than the benchmark" where the portfolio's average ESG score is higher than the benchmark's average ESG score.

This sounds good, but in practice tells us very little about the sustainability of the portfolio.

There's lots of commentary at responsible investment conferences on the lack of correlation between ESG ratings providers.

In May 2022, Tesla was removed from the S&P 500 ESG index. Elon Musk responded exactly as you'd expect: "Exxon is rated top ten best in world for environment, social & governance (ESG) by S&P 500, while Tesla didn't make the list! ESG is a scam. It has been weaponized by phony social justice warriors" (Musk 2022).

The S&P 500 ESG Index includes the largest US companies subject to minimum ESG requirements, including racial discrimination, working conditions and health and safety, all of which affected Tesla's and Exxon's ESG scores.

This is because "ESG" is typically an assessment, relative to peers, of a company's processes, and the extent to which they are subject to ESG risks, and how well those risks are managed. It is not typically an assessment of the sustainability of the company's products.

I have no concerns with this, as long as the limitations of the approach are understood. A portfolio with a "high ESG score" does not mean the portfolio is sustainable (this is something that some policymakers appear to have misunderstood in reporting requirements). ESG scores are an assessment of how well the companies in the portfolio are managing their financially material ESG risks relative to their peers.

The ESG scores provided by data and ratings agencies are largely unregulated, while their use in portfolio construction is considerable. Some regulators are considering requiring ESG data and ratings agencies to publish their methodologies. Even still, if ESG scores are responsible for investment's future, it's a bleak one. ESG scores should be an input into an investment process and a little more.

I expect there will be considerable disruption to the ESG ratings industry in the months ahead, due to increased regulation and AI technology disruption.

References

FRC (2019a), Proposed Revision to the UK Stewardship Code Annex A—Revised UK Stewardship Code. [online]. Available from: https://www.frc.org.uk/getatt achment/bf27581f-c443-4365-ae0a-1487f1388a1b/Annex-A-Stewardship-Code-Jan-2019.pdf (Accessed, June 2023).

FRC (2019b), Feedback Statement: Consulting on a Revised UK Stewardship Code. [online]. Available from: https://www.frc.org.uk/getattachment/291 2476c-d183-46bd-a86e-dfb024f694ad/200206-Feedback-Statement-Consultat ion-on-revised-Stewardship-Code-FINAL-(amended-timetable).pdf   (Accessed, June 2023).

FT Adviser (2021), Schroders 'Frustrated' at Failure to Make Stewardship Code. [online]. Available from: https://www.ftadviser.com/regulation/2021/09/07/sch roders-frustrated-at-failure-to-make-stewardship-code/ (Accessed, January 2023).

Kay (2012), The Kay Review of UK Equity Markets and Long-Term Decision-Making. [online]. Available from: https://assets.publishing.service.gov.uk/govern ment/uploads/system/uploads/attachment_data/file/253454/bis-12-917-kay-rev iew-of-equity-markets-final-report.pdf (Accessed, January 2023).

Musk (2022), Elon Musk. [online]. Available from: https://twitter.com/elonmusk/ status/1526958110023245829?lang=en (Accessed, January 2023).

PRI (2016), US Roadmap. [online]. Available from: https://www.unpri.org/dow nload?ac=4353 (Accessed, January 2023).

6

Early Regulation

The Start of ESG Regulation

Part led by policymaker, part led by investor, also in around 2015–2018, we started to see regulation on responsible investment. In Europe it tended to be policymaker-led, in the US, investor-led, which is simply a function of the policymaking cultures in both regions.

In a demonstration of how quickly regulators have responded to responsible investment, just a few years ago, I could explain every major responsible investment-related regulatory intervention made in all major markets. Now, I struggle to explain the EU's flagship SFDR—just one regulatory intervention, in just one market (albeit, a complicated one). Indeed, the European Commission's sustainable finance team employs around 22—and growing—professionals. While markets remain unsustainable, sustainable finance regulation continues apace.

Keeping on top of regulation is a challenging role for responsible investment professionals.

It is worth taking some time to understand the origins of sustainable finance regulation, because it helps us to understand the current regulatory architecture. Here, I will draw on several PRI publications I authored or co-authored, including the Global Guide to Responsible Investment Regulation.

In what I thought was a big win at the time, I secured funding from MSCI to undertake the "Global Guide" research. It wasn't much, around £20,000. We first researched and then catalogued sustainable finance regulation from around the world. We then sought to develop a typology, which still holds.

© The Author(s), under exclusive license to Springer Nature Switzerland AG 2023
W. Martindale, *Responsible Investment*, https://doi.org/10.1007/978-3-031-44536-1_6

Responsible investment regulation, we found, could be broadly grouped into one of three categories.

1. The first category was pension fund ESG integration requirements. Considering ESG issues was often voluntary, opt in and couched in caveat. The drafting was weak and the terminology was confused. But it started to move ESG from something pension funds can choose to do, to something that, if they choose to do it, they must disclose how they do it, and ultimately, to something they must do. In turn, this raised attention to ESG issues and standards.
2. The second category was stewardship codes. Stewardship codes around the world vary. Some, like the UK's or Hong Kong's Principles of Responsible Ownership were well-drafted and regulatory-led. Others were essentially industry group codes, perhaps well-drafted but little more than a guide and lacking in enforcement.
3. The final category was corporate disclosure, which our research identified to be by far the most prevalent of responsible investment-related regulations. This category varied too. Some regulations were issue specific, for example, disclosures relating to water use or gender pay gap, but some were more comprehensive.

We had a nice line on corporate disclosure, which I continue to quote to this day. Corporate disclosure "is a necessary, but not sufficient condition for supporting responsible investment" (PRI, 2016a). For responsible investment to be successful, we need to mandate companies to disclose ESG issues, but company disclosure alone will not lead to more sustainable markets.

For policymakers, we made the following recommendations: "Policymakers should:

- articulate the role capital markets should play in contributing to a sustainable financial system, with measurable objectives;
- for investor-related regulation:

    build the evidence base on investor practice to understand how capital markets currently contribute to, or undermine, sustainable economies; strengthen policy design – tentative drafting and easy opt-outs mean responsible investment policy is often easy to disregard; improve monitoring and communicate the impact – clarify how regulators' mandates contribute to sustainable economies.

- introduce mandatory corporate reporting on ESG issues. Future climate reporting should aim for international consistency."

The report's dated, but many of the recommendations still hold.

As examples of early ESG regulatory requirements, the Ontario Pension Benefits Act, the UK Pensions Act and the US Department of Labor Employee Retirement Income Security Act (ERISA) are worth reviewing in more detail. Their history and divergence are an interesting microcosm of responsible investment-related regulation the world over.

Earlier iterations of policy change tended to rely on principle-based implementation rules. For example, the 2016 revision of the Ontario Pension Benefits Act states that "a plan's statement of investment policies and procedures is required to include information as to whether ESG factors are incorporated into the plan's investment policies and procedures and, if so, how those factors are incorporated" (Financial Services Commission of Ontario, 2017).

ESG incorporation tended to be optional, and for pension plans that did incorporate ESG issues, the pension plan was required to disclose their policies and procedures to the regulator. While at the time this was considered an important intervention, policymakers in other markets have since provided more precise regulatory requirements to accelerate further market acceptance. At the time, some Canadian pension funds told me that it would potentially limit further adoption of responsible investment policies, because it was now a regulated activity and so incurred additional costs.

The first iteration of the UK ESG-related regulations required pension fund trustees to disclose the extent to which pension schemes incorporate "social, environmental and ethical" factors into account "if at all" when making investment decisions (Pensions Policy Institute, 2018).

Given interpretations of fiduciary duty, and the Cowan vs. Scargill English trusts law case, pension trustees tended to err on the side of caution, and would rarely incorporate issues that were not considered financial.

Even where sustainability issues were considered financial, trustees tended to be sceptical, sustainability issues were badged "non-financial," and proponents would be required to repeatedly provide evidence of their financial materiality.

Because it was not possible to prove, in all instances, that sustainability outperforms, just one instance of sustainability under-performing was seen as a reason not to incorporate sustainability issues in pension scheme decision-making.

In 2018, the UK revised its Occupational Pension Scheme Regulations. The revised regulation stated that "Funds must disclose their policies in relation to financially material considerations. This is defined as including ESG issues and climate change."

In other words, disclosure is no longer optional, and ESG issues, including climate change, are considered financially material considerations for pension schemes. This was a significant change, and in the UK at least, remains the prevailing interpretation.

In the US regulatory agencies are political, led by political appointees, appointed by the president. For the overseas observer this seems dysfunctional, and the evidence suggests, it often is.

A good example is Department of Labor (DoL) ERISA rules on ESG issues, which have flip-flopped back and forth under successive administrations.

Many pension schemes, reluctant to participate in political ping pong track the middle ground, with some sort of reference to ESG issues, but when and only when the evidence on financial materiality is watertight.

As the industry (in Europe and the UK at least) begins to consider net zero and societal outcomes, the US DoL regulation is dated, even under Democrat administrations.

Some investors pushed back against ESG regulation, but most did not. Regulation has many benefits to investors in achieving their responsible investment objectives. It raises the playing field, rewards first-movers, addresses greenwashing and, most importantly, helps evolve our capital markets to be more sustainable. Regulation can also lead to efficiencies, defining terms and processes across the intermediation chain.

In the following sections, I'll take a look at the growth of responsible investment regulation in several countries where I've worked, alongside local consultants, investors and policymakers.

For readers from those countries, the summaries will be high level. But for those not from those countries that work in responsible investment, the summaries should provide some context as to the complex interplay of responsible investment approaches, terminology and disclosure.

I'll start with South Africa.

Focus on South Africa

While much of the responsible investment-related literature references developed markets, emerging markets offer some interesting case studies of ESG-related regulation. Few places more so than South Africa.

In 2015, I spent some time in Johannesburg and Cape Town meeting with South African policymakers, investors and stakeholder groups. Following a few months of research, the PRI launched a South African Fiduciary Duty in the 21st Century roadmap at a hotel in Greenmarket Square, Cape Town, in October 2015.

The panelists for the discussion were South African representatives. Our interviews comprised a mix of black and white South Africans.

But the first question was, "how do you feel about the white asset management sector extracting high fees for ESG products from predominantly black pension funds?"

I should have been prepared for such a question, but of course I was not— who was I to answer such a question? Other panelists kicked in, but the event didn't recover, and the launch was largely unsuccessful. It was an important reminder that in countries like South Africa, Brazil and India, "ESG" is very abstract and often an unimportant concept.

South Africa has a range of unique social and environmental issues.

At the time of writing, unemployment in South Africa is around 30%. Over 50% of under 35s are unemployed, of which 90% are black. Many black South Africans who are unemployed lack the education for the modern workplace, rooted in apartheid exclusions.

There is also considerable spatial injustice—smaller towns are a labour source for wealthier cities. The black poor were moved systematically out of places of opportunity. And so black economic empowerment is a social and economic necessity.

South Africa is a coal economy at the mercy of its monopoly provider, Eskom. Eskom is an integrated monopoly utility. Eskom has huge existential, environmental and financial issues.

Coal provides a large part of South Africa's electricity and employment. Given unemployment levels, politicians are under pressure to retain coal-related employment and as such prop up the coal industry: "We've got all these miners – we need to make sure they keep their jobs."

South Africa's powerful labour unions with deep political influence will choose coal every time. Eskom has been badly run for years, providing "free" electricity to political connections, and as such huge political and vested interest challenges.

More recently, there are regular blackouts. For a country with superior solar resource (one of the top three in the world, given the country's space and land), there remains considerable coal incumbency.

The people who suffer most from the consequences of pollution and climate change-related weather events are the poor—and that's particularly the case in South Africa. Poorer South Africans tend to live in "down wind" areas and in areas close to fossil fuel production. Fossil fuels are used by the rich and the externalities are borne by the poor.

South Africa's national development plan includes decarbonisation of the South African economy, but the exclusionary nature of apartheid has led to a trust deficit between the state and private sector.

The solar developers are coming to market. The new energy technologies do two things—provide people with local, cheaper electricity, and over time, push coal out of the market.

But the political discourse is wrong. The approach is: "We've got coal in the ground, and that's an asset", but South Africa has also "got sun in the sky, and that's an asset too."

As the world decarbonises, South Africa does not want to be the last country to transition its energy systems to new renewable technology. If South African policymakers embraced solar, it would almost certainly become an export to neighbouring African economies.

It is important to place our understanding of South Africa's responsible investment-related regulation against this backdrop.

South Africa's pension market is similar to the UK's or Canada's. A few large schemes, but a long tail of small schemes with weak governance.

South Africa's regulatory frameworks on responsible investment are well developed.

Regulation 28, overseen by the Financial Services Board, requires pension funds to "support the adoption of a responsible investment approach … including factors of an environmental, social and governance character."

When Reg 28 was introduced, it was ahead of most other capital markets. Reg 28 also requires pension funds to include B-BBEE as part of their service provider selection process.

B-BBEE stands for Broad-Based Black Economic Empowerment, or BEE for short. It is a government policy to advance economic transformation and enhance the economic participation of black people in the South African economy (Norton Rose, 2018).

BEE requires pension funds to invest in companies with strong black empowerment performance (against a scorecard overseen by policymakers).

For obvious reasons this is unique to South Africa's tragic history. To some extent, other governments have followed suit, seeking to route private capital to social goals.

In 2021, the UK Johnson government tried to do this with their economic "big bang". As with much of the Johnson government's policymaking, it was chaotic, writing to pension schemes offering them dinner with Prime Minister Johnson if they committed investments to UK projects.

South Africa's BEE policies have been widely criticised. Power remains concentrated, unemployment and inequality remains unacceptably high. But the Reg 28 pension regulations are on the whole well developed.

Alongside Reg 28 is a principle-based code, "CRISA" (Code for Responsible Investment in South Africa). It is not resourced and signatories to the Code are not assessed, but it is well-written and has normative influence.

Accompanying CRISA is the King Code, its corporate governance equivalent, overseen by the Institute of Directors.

The country's largest pension scheme, the Government Employees Pension Fund (GEPF), with over 125 billion dollars (USDs) in assets under management, and PIC, Public Investment Corporation, Africa's largest asset manager with 145 billion dollars (USDs) in assets under management, set the tone for the rest of the institutional investment market, in South Africa, and arguably, Sub-Saharan Africa.

A challenge in South Africa is regulatory capacity. Given the volume of pension funds, insiders told me that regulatory oversight is inadequate.

But perhaps the biggest challenge is the environmental, social and political backdrop. Responsible investment has a role to play in South Africa's transition, but it is largely a backseat role, supporting and enabling, but not necessarily leading—that's in the hands of politicians, Eskom and the labour unions.

Focus on Brazil

In some respects, Brazil is similar to South Africa. It is the powerhouse economy of its continent. It has a legacy of interest and action on responsible investment topics.

The pensions regulator, PREVIC (which loosely stands for the superintendence of private pension funds) issued a resolution in 2009, which required pension funds to comply with associated governance requirements, including whether or not they take into account environmental and social issues.

Subsequent resolutions in 2018, 2019 and 2020 gradually started to strengthen ESG-related requirements.

"ESG aspects should be considered when analyzing investment risks, whenever possible" and again "The investment policy statements must contain guidelines for compliance with ESG principles, preferably, by sector of economic activity" (Brazil Government, 2021).

For its part, the Central Bank requires the financial sector companies it regulates to publish a sustainability policy, although this is typically a Corporate Social Responsibility (CSR) policy, with little scrutiny of the content of the policy (PRI, 2017a).

There is a stewardship code, originally prepared by AMEC (the Brazilian association of capital market investors) and refreshed in 2021 in partnership with CFA Brazil (Chartered Financial Analyst Brazil). The stewardship code includes seven principles, which are succinct and clear. The first principle is to implement and disclose a stewardship programme. Each principle is accompanied by a paragraph or two of guidance.

As with South Africa, the issue is not the drafting, but the implementation. Whatever the market, stewardship suffers from the "tragedy of the commons", and without enforcement that will not change. Market incentives alone—in Brazil and elsewhere—are insufficient. Not to take away from AMEC and CFA's work here, but AMEC is not the right institution to "own" the stewardship code, rather this should be CVM (capital markets regulator) or PREVIC, the regulators.

Undoubtedly ESG integration efforts seem negligible compared to the gravity of Amazon deforestation. The Investor Policy Dialogue on Deforestation (IPDD) is a collaborative initiative to engage "public agencies and industry associations" across a number of countries, including Brazil, on the topic of deforestation (Storebrand, 2022).

The initiative's secretariat support is provided by the Tropical Forest Alliance (TFA), an initiative hosted by the World Economic Forum, and supported by the PRI.

Around 60 investors have subscribed to the initiative. The Brazil working group was the IPDD's first. The IPDD set five objectives (Storebrand, 2022). The outcomes expected by the investors are:

1. Significantly reduce deforestation rates
2. Enforce Brazil's Forest Code
3. Reinforce Brazil's agencies tasked with implementing environmental and human rights legislation, and avoid any legislative developments that may negatively impact forest protection
4. Prevent fires in or near forest areas to avoid those as seen in 2019 and 2020
5. Allow public access to data on deforestation, forest cover, tenure and traceability of commodity supply chains.

The rationale for engaging is twofold. The risks relevant to companies operating in the Amazon (or companies buying from companies operating in the Amazon) include reputation, regulation and litigation risk. And the systemic risks caused by billions of tons of carbon dioxide released into the atmosphere through forest fires and reduction in carbon sequestration as the Amazon shrinks.

President Lula's election in October 2022 was expected to slow deforestation. Perhaps his victory gives space to policymakers and investors to further advance responsible investment regulation in Brazil too.

Focus on Australia

Australia was at the forefront of responsible investment. Its large, well-resourced superannuation schemes were quick to understand the importance of responsible investment, to integrate ESG issues in investment decision-making, to undertake stewardship and to consider impact.

Cbus, the Construction and building unions superannuation, was one of the first—and one of the few—to undertake SDG reporting. Cbus and HESTA, the Health Employees Superannuation Trust Australia, occupied successive PRI board positions.

Responsible investment in Australia benefits from well-resourced, successful industry groups, such as ACSI, the Australian Council for Superannuation Investors and AIST, the Australian Institute of Superannuation Trustees. Both ACSI and AIST undertake high quality work programmes on responsible investment topics.

Prior to 2015, Australia was the responsible investment global leader.

At the time, Australia's regulation was largely consistent with its peers. The prudential regulatory authority (APRA), which oversees the superannuation industry, undertakes a principles-based approach to regulation. In 2013, in response to a super system review, known as the "Cooper review", APRA updated regulation to address ESG issues (PRI, 2017d).

The language however was confused (APRA, 2013). Para 34 of what's called SPG 530 said:

"An ethical investment option is typically characterised by an added focus on environmental, sustainability, social and governance (ESG) considerations, or integrates such considerations into the formulation of the investment strategy and supporting analysis."

However, in most markets this is not the case. ESG considerations are more commonly understood to be financial, not ethical.

Para 36 of SPG 530 said:

"In offering such investment options, a prudent RSE licensee would be mindful of exposing the interests of beneficiaries to undue risk stemming from matters such as a lack of diversification, where investment in some industries are excluded or a positive weighting is placed on certain non-financial factors as a result of ESG considerations."

Again, ESG considerations are widely interpreted to be financially material and so the text does not make sense. That said, the confused language was not uncommon at the time, similar to the UK's Pension Act, US Employee Retirement Income Security Act and the EU's occupational pension funds directive, known as IORP.

Australia has a stewardship code, but it is industry led. FSC, the Financial Services Council, introduced non-binding guidance known as the "blue book" (PRI, 2017d). As with other industry codes, the code is well-drafted, and as an industry group, FSC has influence. But it is not the same level of influence as a regulator.

I spent some time in Australia in 2016 to write the Fiduciary Duty in the 21st Century Australia roadmap. I attended a meeting with APRA, accompanied by an Australian member of the PRI's Australian committee (a working group that supported the PRI on topics relevant to Australia).

Afterwards he said to me, "you got the right balance of telling them they're backward without telling them they're backward." That wasn't my intention and indeed, following my visit, there was at least one complaint from Australian signatories about my approach.

But given Australia's size, wealth, well-funded, well-governed superannuation schemes, its sun and wind capabilities and great quantities of natural resources, including uranium for nuclear—and small, well-educated population—I struggled to understand why Australia was not the world's climate leader, and why Australian signatories seemed unprepared to engage with policymakers in ways we were seeing in the UK and Europe. I think, as with the US, I'd misunderstood how politicised these issues had become.

Over the past decade or so, Australia has had its fair share of populist prime ministers unwilling to accept climate change. In particular, Tony Abbott from 2013 to 2015, who, in a speech in London in 2020, said climate change is "probably doing good" (The Guardian, 2017a), as well as, Malcolm Turnbull from 2015 to 2018 and Scott Morrison from 2018 to 2022. Morrison brought coal into the Australian Parliament saying coal would "keep the lights on" (The Guardian, 2017b).

Perhaps it was inevitable that, on responsible investment, Australia would stand still or indeed, regress, overtaken by Europe and the UK.

A number of Australians working in responsible investment found their way to the UK. Not just at PRI, but the Climate Bonds Initiative, and several asset managers.

In addition to politics, there are a few quirks of regulation that give rise to short-termism. MySuper rankings compare investor performance on an annual basis across three, five and seven-year time periods. A recent refresh of the Australia roadmap concluded that, "asset owners may be disincentivised from addressing sustainability outcomes and mitigating system-level risks as the financial benefits of doing so are likely to be realised only in the long term" (PRI, 2022).

Australia is an example of where politics matters. Australia is an asset owner-led financial system with trillions of dollars in long-term pension savings. Australian superannuation funds are well-resourced, well-governed, and were quick to understand the benefits of responsible investment. But from 2015 to 2022, progress stalled. In short, responsible investment was held back by politics.

With the election of Antony Albanese in May 2022, change was underway.

In November 2022, APRA published a draft prudential practice guide with much clearer language (APRA, 2022).

"APRA expects an RSE licensee would demonstrate an understanding of the risk and opportunities present in a range of ESG factors, and the extent to which they may have a material impact on the financial risk–return profile of the RSE licensee's investment portfolio, including an assessment of climate risk exposures."

"An RSE licensee would demonstrate how ESG risk considerations are integrated into investment analysis, decision-making and oversight, ensuring that the appropriate resources are available to identify and respond to material ESG factors. Where an RSE licensee determines that ESG risks are financially material, it would demonstrate how this is reflected in the investment strategy and document how the risks will be managed."

The language closely follows the UK Pensions Act, including, "An RSE licensee may pursue additional objectives from investments, such as environmental or social impacts, where it can demonstrate that pursuing such additional objectives is consistent with the delivery of investment outcomes to members."

This was a step forward. I expect it's likely there remains some reticence from decision-makers on real-world sustainability impact.

In November 2022, the Australian Sustainable Finance Institute (ASFI), an industry-led initiative with members ranging from PWC and EY to State Super and Aware Super, called on Australia to join the EU-led International Platform on Sustainable Finance (IPSF), which, among other things, seeks to harmonise Taxonomies.

In December 2022, the Australia Treasury, led by Jim Chalmers, launched a consultation on climate-related financial disclosures (Australia Treasury, 2022). The 19 questions were expected to provide a springboard for additional climate reporting, likely based on International Sustainability Standards Board (ISSB) disclosure proposals, including reporting Scope 3 GHG emissions for some sectors.

In the coming years, Australia will be the country to watch (PRI, 2022). On the one hand, Australia's trade relations with China and other Asian markets, in particular, including coal supply, its extensive mining operations and fossil fuel resources may hold it back. On the other hand, its wealth, well-resourced superannuation funds, and for the first time in a generation, progressive politics may propel it back in to pole position when it comes to responsible investment.

Focus on Japan

Japan's government pension fund (GPIF) is the world's largest asset owner. What GPIF says matters, not just in Japan, but globally.

For several years, Hiro Mizuno was GPIF's CIO. A government appointee, Mizuno joined PRI's board and committed GPIF to a number of responsible investment topics. He even, he once told me, lobbied US regulators on responsible investment topics during the Trump administration, such was his influence.

GPIF is not and is unlikely to be the highest conviction responsible investor, but that should not take away from Japan's recent progress, and GPIF's starring role.

Mizuno once described Japan as the "desert of ESG". Japan's financial sector is cautious, possibly even, conservative, and GPIF will stay within what's politically acceptable. It didn't stop Mizuno pushing as hard as his mandate allowed. Mizuno's term concluded in March 2020. He's since been appointed as the United Nations special envoy on Innovative Finance and Sustainable Investments, and has joined the boards of Tesla and Danone.

GPIF's conviction on responsible investment has its roots in Abenomics (PRI, 2017b). The Japan revitalisation strategy was approved by the Cabinet in June 2013. It followed 20 years of economic sluggishness. Abenomics sought to grow the Japanese economy and spur capital investment.

The strategy set out expectations on corporate governance and stewardship, specifying "preparation of principles (a Japanese version of the Stewardship Code) for institutional investors in order to fulfill their stewardship responsibilities, such as promoting the mid- to long-term growth of companies through dialogue" (PRI, 2017a, 2017b, 2017c, 2017d).

In part, the codes were intended to support investors to shift away from low risk, low return fixed income investments to higher risk, higher return equity investments—and in doing so, spur growth. As the FSA's strategy review says, "This reflects the FSA's overriding mandate which expressly seeks to promote the 'sustainable growth of business activities and the wider economy' in Japan and to cope with 'uncertainties in the global economy' including arising from technological change" (PRI, 2017b).

The codes have been a success, with improvements in corporate governance (in a corporate culture that Japanese investors told me was not particularly meritocratic), which, in addition to domestic investors, has opened Japan to overseas investors.

The codes have undergone successive revisions. Adherence is not scrutinised in the same way as the UK code. I expect that in Japanese culture, regulatory guidance is more readily adopted than, say US or European culture. In my experience, most commentators hail the role they've played in the Japanese economy.

More recently, Japanese regulators have, tentatively, approached TCFD. The stock exchange has provided training on TCFD reporting and started to introduce guidance for TCFD reporting. Progress is, however, frustratingly slow. TCFD reporting is not yet a requirement (JPX, 2022).

The Japanese government has set a target to reduce emissions by 46% with a 36–38% share of renewables in the power mix, and Japan has committed to net zero by 2050 (Reuters, 2021). Nuclear remains a political challenge, with over 200,000 people displaced over a decade after the March 2010 earthquake and the resulting Fukushima Daiichi nuclear powerplant disaster. Offshore wind is the major growth area, but coal remains the base-load supplier.

I expect there will be further attention to responsible investment in Japan in the months ahead, likely green bond issuances from both government and major Japanese companies, as well as the introduction of mandatory TCFD reporting.

Japan is however unlikely to be a responsible investment leader and despite its economic influence, is often overlooked in favour of developments in China.

Focus on China

I've visited China a few times, staying in Shanghai and Beijing, and of course, Hong Kong. My most recent visit, now a few years back, was in June 2019.

After a day's work, I returned to my Beijing hotel room to find it being searched. The two people exited my room, but did not rush to do so. I have no idea why my room was being searched, nor by whom. But almost certainly, whoever it was that searched my room wanted me to know about it.

As part of our visit, we had undertaken a series of meetings with the People's Bank of China, the Ministry of Human Resources and Social Security, the stock exchanges and investors, and perhaps that was enough to trigger a room search.

But I can't imagine my work was of interest. Indeed, PRI was particularly cautious in its approach, framing recommendations in the conditional tense, "the MoHRSS could introduce a stewardship code", and mostly awareness raising (PRI, 2018). While the European-based Fiduciary Duty in the 21st Century project team wanted to go further, my Chinese colleagues helped to ensure that the recommendations were politically and culturally acceptable.

There's an argument that responsible investment, as a soft-form of influence for change, is less impactful in state-run economies. If the government chooses change, then change will happen. The influence of private investors is less significant than non-state-run economies.

During my visits, I only secured one meeting with China's state pension fund, the National Social Security Fund (NSSF). As with GEPF in South Africa and GIPF in Japan, NSSF sets the tone for investment decision-making for the Chinese pension industry. NSSF is not a signatory to PRI, and joining would not be straightforward. Investment decisions are subject to oversight from government officials and could not be seen to be subject to overseas interference.

At time of writing, there are four asset owner signatories in China, the best known (and first to sign) is Ping An Insurance, joining PRI in 2019.

As is the case the world over, the success of responsible investment is tied to individuals, and in China, it was Dr. Ma Jun, who was a member of the PBOC's monetary policy committee from 2018 to 2020.

Dr. Ma drafted China's green finance guidelines in 2015–2016 and, as well as leading or contributing to a range of subsequent green finance initiatives, is a well-known advocate for China's approach to green finance at responsible investment conferences the world over.

China has committed to net zero GHG emissions by 2060, with emissions peaking by 2030. China's transition is, and will be, in part financed by green bonds.

China is a green bond leader, rapidly expanding its green bond issuances in 2021 to over 100 billion USDs, although, there are differences in how China—and overseas initiatives, such as the Climate Bonds Initiative—determine what constitutes a green bond.

According to CBI, in 2021, the majority of green bond issuances were "non-financial corporates" (58%) of which 97% were issued by state-owned enterprises, of which, most were invested in renewable energy, low carbon transport and low carbon buildings (CBI, 2022).

As China further opens its economy to overseas investment, informed by President Xi's Belt and Road Initiative, which seeks to increase integration between China's economy and that of its trading partners, green finance, environmental disclosure, and perhaps even, TCFD or taxonomy reporting, is likely to be a feature of China's capital markets.

Overseas responsible investors can signal their support through investment. The role of overseas investors in the stewardship of Chinese companies is much more challenging. And while I think it does make sense for organisations such as PRI or UNEP FI to have Chinese-based staff, their role is more politically sensitive than their overseas peers.

Focus on Europe

When it comes to responsible investment, Europe is the world's leader. But that's very recent history.

As recently as 2014, Europe's environment ministry commissioned EY to undertake a study into whether institutional investors could integrate environmental factors into their investment policies and decision-making processes (DG Environment, 2014).

As with every study hitherto and hence, the answer was "as long as it is relevant to financial returns". Of course, as we discussed earlier, it depends on what time frame, and how it's measured (stock-specific or systemic risks).

At the time, the IORP II directive, which stands for "Institutions for Occupational Retirement Provisions", in short, workplace pension funds, did not include provisions for ESG integration.

Again, as recently as December 2016, the EU updated the IORP directive to include ESG integration requirements (European Commission, 2016). "The system of governance [of pension funds] shall include consideration of environmental, social and governance factors related to investment assets in investment decisions, and shall be subject to regular internal review."

As a demonstration of our progress on responsible investment, responsible investment group ShareAction, a more radical group than PRI, was "delighted" calling the regulation a "key litmus test on EU policymakers' appetite for action on building a sustainable economy" (IPE, 2016). I don't think ShareAction would say that now.

We'll discuss the more recent EU policy interventions later, but at the time, the other two responsible investment-related policies were the "Non-Financial Reporting Directive" (NFRD) and the "Shareholder Rights Directive" (SRD). The terminology "non-financial" is clearly unhelpful, although the criticism is easy to dismiss.

Firstly, the terminology was a hangover from previous EU policymaking, and changing the name of the directive was politically unnecessary. Secondly, as a senior investor once said to me, "it is non 'traditional' financial. ESG issues have not traditionally been part of financial accounts, hence 'non-financial'." In other words, we shouldn't worry about it.

The NFRD, which came into effect in 2018, requires companies to disclose environmental, social, employee, human rights, anti-corruption and bribery and diversity issues.

It is the precursor to the Corporate Sustainability Reporting Directive (CSRD), first proposed by the European Commission in April 2021, which, among other requirements, introduces "double materiality" disclosures to companies.

The Shareholder Rights Directive (SRD) II was introduced in 2017, an amendment to SRD I, which was introduced in 2007. SRD applies to companies with a registered office in the EU and listed on an exchange in the EU. The core requirements of SRD II are a transparency exercise, enabling companies to know who their shareholders are. It also requires European investors to monitor investee companies, including "social and environmental impact and corporate governance".

On a comply or explain basis, investors must disclose annually how their engagement policy has been implemented, their voting behaviour, their use of proxy advisors, and significant votes.

SRD II is not nearly as comprehensive as the UK Stewardship Code. But as a baseline for shareholder engagement, it remains an important intervention.

Prior to the launch of the EU's High Level Expert Group (HLEG), EU responsible investment-related regulation was typical, in other words, it was no more advanced than the UK's, Canada's or Australia's. It comprised ESG disclosure requirements for companies, a nod to stewardship and ESG considerations for workplace pensions.

But with HLEG that all changed. "Expert Groups" are not uncommon in the EU. But few attracted the fanfare of the HLEG on sustainable finance. Launched in December 2016, the HLEG's mandate was time limited. You have a mandate, a year and some (limited) secretariat support to set out recommendations for a "comprehensive European strategy on sustainable finance" (European Commission, 2017).

Six of the 20 experts were British, reflecting the UK's contribution to sustainable finance in Europe. The chair was the very impressive Christian Thimann, a German working for the French firm, AXA.

Within just seven months, the HLEG published an interim report, subject to consultation (HLEG, 2018). There were 300 consultation responses.

The HLEG group worked well. The participants were highly committed, attending meetings as often as once a month, supported by their employer (it was not a paid role). It was the right group of individuals, at the right time, supported by European Commission staff, and groups like PRI. The meetings were well-tempered, engaging and constructive. The HLEG divided into subgroups to develop the recommendations more fully, including a short offsite.

The final report followed in January 2018 setting out eight "key recommendations", eight "cross-cutting recommendations", eight recommendations for "financial institutions" and four thematic recommendations. At exactly 100 pages, it represented perhaps the biggest shift in thinking in responsible investment's short history (HLEG, 2019).

It was and remains a terrific piece of work and huge credit is due to the policymakers and experts. For responsible investment professionals, it's core reading, whichever your jurisdiction. At the helm was Valdis Dombrovskis, the European Commission's vice president for the Capital Markets Union.

I didn't get to know Dombrovskis, who was the former prime minister of Latvia from 2009 to 2014. In the few times we did interact, he made it clear to me that he was not one for small talk. But he did (at least it seemed to me) believe in the potential of sustainable finance, giving the HLEG his full political backing. At the time, there was resistance to his sustainable finance plans, particularly from member states. But Dombrovskis shot back, saying in a major speech in October 2017, "we need to increase the private capital flow to sustainable finance" (Dombrovskis, 2017).

His mandate clearly came from the top, "The financial sector will have to throw its full weight behind the fight against climate change", said Jean-Claude Juncker in 2018, as then-president of the European Commission (Juncker, 2018).

European policymakers have, on the whole, adopted the HLEG's recommendations. It's perhaps the best example of policymaker and industry collaboration on sustainable finance. We'll take a look at the EU Taxonomy, the SFDR and the Technical Expert Group (TEG) later.

I asked Nick Robins about his involvement in HLEG. "It was exciting. My recollection of sequencing of events is as follows. Clearly you had the Paris Climate Agreement, but also China's G20 initiative on green finance. There was almost a race, a competition, and the EU needed to up its game which resulted in the HLEG."

"It's extraordinary how much consensus we got. In Christian Thimann, we had this fantastic European statesman, a public servant."

"One interesting area of contention was was the role of the green supporting factor. The Commissioner wanted the green supporting factor to reduce the capital weights in the Basel rules for sustainable assets. But most of HLEG said no, it's not the right way to encourage investment in clean energy assets as you're muddling a risk tool with a promotional goal."

"We also thought some of the things we proposed would be straight forward, such as the taxonomy."

"Of course, there was a huge personal thing of being a Brit. Quite a few were Brits. After Brexit, the UK influence on trends in sustainable finance I think has taken a hit."

"When I started as head of unit of the Capital Markets Union," Martin Spolc told me, "my predecessor told me, 'sustainable finance, you should watch it. It's going to be big'."

"He said, 'you really need to meet the HLEG members. They are really good. You need to have a conversation with each of them so that you understand where they are coming from'."

"At the time, the focus for me was banking regulation and how to strengthen capital markets in Europe. We did have some work underway on long-term investment, but we didn't really talk about sustainable investment."

"My first introduction to sustainable investment was around the end of 2017. Sustainable finance was just one of many points on a long list of actions for the CMU."

"But it was only after I attended one of the last HLEG meetings when I looked at the draft report that Christian Thimann was putting together, and I realised, 'wow', there is something in this."

"And the more I started to look into it, the more I realised, we can make something tangible and meaningful out of this."

References

APRA (2013), Prudential Practice Guide, SPG 530 – Investment Governance. [online]. Available from: https://www.apra.gov.au/sites/default/files/prudential-practice-guide-spg-530-investment-governance_0.pdf (Accessed, January 2023).

APRA (2022), Prudential Practice Guide, Draft SPG 530 Investment Governance. [online]. Available from: https://www.apra.gov.au/sites/default/files/2022-11/Prudential%20Practice%20Guide%20-%20Draft%20SPG%20530%20Investment%20Governance.pdf (Accessed, January 2023).

Australia Treasury (2022), Climate-related financial disclosure. [online]. Available from: https://treasury.gov.au/consultation/c2022-314397 (Accessed, January 2023).

Brazil Government (2021), Pension Fund Managers' Use of ESG Criteria. [online]. Available from: https://www.gov.br/previc/pt-br/publicacoes/pesquisa-asg/results-of-a-survey-questionnaire-answered-by-brazilian-pension-funds/results-of-a-survey-questionnaire-answered-by-brazilian-pension-funds.pdf.

Climate Bonds Initiative (2022), China Green Bond Market Report 2021. [online]. Available from: https://www.climatebonds.net/resources/reports/china-green-bond-market-report-2021 (Accessed, January 2023).

DG Environment (2014), Resource Efficiency and Fiduciary Duties of Investors. [online]. Available from: https://ec.europa.eu/environment/enveco/resource_efficiency/pdf/FiduciaryDuties.pdf (Accessed, January 2023).

Dombrovskis (2017), EU Monitor [online]. Available from: https://www.eumonitor.eu/9353000/1/j9vvik7m1c3gyxp/vkicj1n00wxa?ctx=vhyzmvnvbbzs&v=1&tab=1&start_tab1=110 (Accessed, January 2023).

European Commission (2016), IORP Directive. [online]. Available from: https://eur-lex.europa.eu/legal-content/EN/TXT/PDF/?uri=CELEX:32016L2341&from=EN (Accessed, January 2023).

European Commission (2017), HLEG Press Release. [online]. Available from: https://finance.ec.europa.eu/system/files/2017-04/161028-press-release_en.pdf (Accessed, January 2023).

Financial Services Commission of Ontario (2017), Statement of Investment Policies and Procedures (SIPP). [online]. Available from: https://www.fsco.gov.on.ca/en/pensions/archives/Pages/sipp-archive.aspx (Accessed, February 2023).

The Guardian (2017), Tony Abbott says climate change is 'probably doing good'. [online]. Available from: https://www.theguardian.com/australia-news/2017/oct/10/tony-abbott-says-climate-change-is-probably-doing-good (Accessed, January 2023).

The Guardian (2017b), Barnaby Joyce juggles coal. [online]. Available from: https://www.theguardian.com/australia-news/live/2017/feb/09/bill-shorten-malcolm-turnbull-union-liberal-labor-politics-live (Accessed, January 2023).

HLEG (2018), Financing a Sustainable European Economy Interim Report. [online]. Available from: https://finance.ec.europa.eu/system/files/2017-07/170713-sustainable-finance-report_en.pdf (Accessed, January 2023).

HLEG (2019), Financing a Sustainable European Economy Final Report. [online]. Available from: https://finance.ec.europa.eu/system/files/2018-01/180131-sustainable-finance-final-report_en.pdf (Accessed, January 2023).

IPE (2016), ShareAction welcomes IORP II focus on ESG risks, stranded assets. [online]. Available from: https://www.ipe.com/shareaction-welcomes-iorp-ii-focus-on-esg-risks-stranded-assets/10014012.article (Accessed, January 2023).

JPX (2022), Environmental Information (TCFD Disclosure). [online]. Available from: https://www.jpx.co.jp/english/corporate/sustainability/jpx-esg/environment/index.html (Accessed, January 2023).

Juncker (2018), Sustainable Finance: High-Level Conference kicks EU's strategy for greener and cleaner economy into high gear. [online]. Available from: https://ec.europa.eu/commission/presscorner/detail/en/IP_18_2381 (Accessed, January 2023).

Norton Rose (2018), Broad-based black economic empowerment – basic principles.[online]. Available from: https://www.nortonrosefulbright.com/en/knowledge/publications/fe87cd48/broad-based-black-economic-empowerment---basic-principles (Accessed, January 2023).

Pensions Policy Institute (2018), ESG: past, present and future. [online]. Available from: https://www.pensionspolicyinstitute.org.uk/media/2398/20181002-ppi-esg-past-present-and-future-report-final.pdf (Accessed, January 2023).

PRI (2016a), Global Guide to Responsible Investment Regulation. [online]. Available from: https://www.unpri.org/policy/global-guide-to-responsible-investment-regulation/207.article (Accessed, January 2023).

PRI (2017a), Fiduciary Duty in the 21st Century Brazil roadmap. [online]. Available from: https://www.unpri.org/download?ac=1386 (Accessed, January 2023).

PRI (2017b), Fiduciary Duty in the 21st Century Japan roadmap. [online]. Available from: https://www.unpri.org/download?ac=1389 (Accessed, January 2023).

PRI (2017c), Fiduciary Duty in the 21st Century South Africa roadmap. [online]. Available from: https://www.unpri.org/download?ac=1390 (Accessed, January 2023).

PRI (2017d), Fiduciary Duty in the 21st Century Australia roadmap. [online]. Available from: https://www.unpri.org/download?ac=1385 (Accessed, January 2023).

PRI (2018), Fiduciary Duty in the 21st Century China roadmap. [online]. Available from: https://www.unpri.org/download?ac=4496 (Accessed, January 2023).

PRI (2022), Legal Framework for Impact Australia [online]. Available from: https://www.unpri.org/download?ac=16940 (Accessed, January 2023).

Reuters (2021), Japan aims for 36–38% of energy to come from renewables by 2030. [online]. Available from: https://www.reuters.com/business/energy/japan-aims-36-38-energy-come-renewables-by-2030-2021-10-22/ (Accessed, January 2023).

Storebrand (2022), Leading the Investors Policy Dialogue on Deforestation [online]. https://www.unpri.org/active-ownership-20/storebrand-asset-management-leading-the-investors-policy-dialogue-on-deforestation/9980.article (Accessed, January 2023).

7

The Growth of Groups

Coopetition

"There are lots of new initiatives with lots of new acronyms" is a concern shared by most responsible investment professionals.

The groups do have their own place, there's a reason they exist, but there's also quite a bit of overlap, giving rise to a new buzzword (that would rile Hemingway): "coopetition".

At times, I would find the relationship between these groups competitive (the interactions between the groups working on Climate Action 100+ or the net zero initiatives were not always constructive), but there is evidently more work to do than groups to do it. The groups do, to an extent, promote each other's work and avoid duplication.

But I often think, imagine if the groups were pooled, same HR, same operations, efficiencies across work programmes, and perhaps better paid staff.

As with any nascent topic, there tends to be a proliferation of stakeholder groups followed by a consolidation. In responsible investment, I think we're still on the ascent.

I asked Steve Waygood about the proliferation of responsible investment groups.

"Some people complain there are too many 'initiatives'. My complaint is that there's not enough 'initiative'." In other words, let's get on with it. That's my view too, and I expect in a few years time, we will start to see the groups align, as we have with the corporate reporting initiatives.

© The Author(s), under exclusive license to Springer Nature Switzerland AG 2023
W. Martindale, *Responsible Investment*, https://doi.org/10.1007/978-3-031-44536-1_7

This section takes a look at some of the groups that I've worked with, or have affected my thinking on responsible investment, their role, how I think investors can get the best from the groups and where relevant, the groups' futures. The order loosely follows chronology.

UNEP FI

The UN Environment Programme Finance Initiative has tenacity. It is one of the longest-running sustainable finance groups.

As a UN agency, unlike PRI, which is "UN supported", UNEP FI must navigate political complexities that other sustainable finance groups can largely ignore. On the other hand, UN status gives UNEP FI access and gravitas. UNEP FI is headquartered in Geneva, with staff across the globe.

UNEP FI incubated PRI. UNEP FI has a non-voting PRI board position, but PRI is independent. UNEP FI also incubated PSI, the Principles for Sustainable Insurance (it's not relevant that investment is "responsible" and insurance is "sustainable", rather a reflection of which word was most widely used at the time of its launch).

PSI is well-known within the insurance industry, but doesn't have PRI's set up. Largely, it undertakes industry-led events, working groups and reports. Unlike PRI, PSI remains within UNEP FI.

More recently, UNEP FI launched the PRB, the Principles for Responsible Banking (we're back to "responsible").

Between PSI and PRB, UNEP FI also launched the Principles for Positive Impact Finance, which are well-intentioned and well-written, but not at all well known. The Principles for Positive Impact Finance were however timely, helping to conceptually develop the way in which financial institutions think about double materiality, even if, the Principles themselves are very much in the background.

The PRB represents perhaps the last part of the financial system to have its own set of "responsible" principles, although, arguably sustainable finance started with the Equator Principles, which are first and foremost for banks to manage the environmental and social risks in their lending activities.

Unlike PRI, PRB incorporates impact in Principle 1, "alignment", and Principle 2, "impact and target setting": "We will align our business strategy to be consistent with and contribute to individuals' needs and society's goals, as expressed in the Sustainable Development Goals, the Paris Climate Agreement and relevant national and regional frameworks." "We will continuously increase our positive impacts while reducing our negative impacts on … people and the environment."

Citi Group and Goldman Sachs are signatories, although I wonder whether their CEOs are aware of their membership—a reflection both of PRB, but also the structure and culture at the banks (and perhaps, the extent to which these issues are often not elevated within banks' hierarchy structures). Indeed, at a US House Committee hearing in September 2022, Citi's CEO said that Citigroup would continue to invest in new fossil fuels (Washington Examiner 2022), appearing to contradict, if not the letter, then the spirit of the Principles.

UNEP FI's multi-year, multi-topic work on natural capital, REDD (Reducing Emissions from Deforestation and Forest Degradation), fiduciary duties, its support for PRI, PSI and PRB, within a complex and politically challenging ecosystem has enabled responsible investment.

A UNEP spin off was UNEP Inquiry, an initiative that, by design, has been and gone, kicking off in 2014, and wrapping up around 5 years later. It was intended as an intervention, and while it lasted longer than its initial mandate, it was not intended to go on and on.

UNEP Inquiry's full name was the (not particularly snappy) United Nations Environment Programme Inquiry into the Design of a Sustainable Financial System.

It was led by Nick Robins and Simon Zadek. Robins had spent four years at HSBC, as head of the climate change centre of excellence.

UNEP Inquiry worked by writing reports, diving into a country, institution or sector, identifying changes and writing about them.

I can't remember the date, but it was warm and sunny. UNEP Inquiry hosted an event at Somerset House, London. About 40 investors were in attendance. After a day of deliberations, one of the conclusions was, "we need a better PRI". Seemingly everyone in attendance agreed we need a better PRI. Although no one agreed what exactly we meant by "better".

The Inquiry was largely dismissive of industry groups, perhaps not without reason, but it was a shame, because the Inquiry team had a good network, and its work was well done. In my experience, the shelf-life of most of its publications was limited. This was deliberate. It was a short, sharp intervention, intended to operate outside of the political challenges that face UNEP FI. I do think there could have been more focus on finding "homes" for UNEP Inquiry's work to implement the report's recommendations.

Its final report, "Making Waves" is an overview of its successes, complete with forewords from the (then) Bank of England governor, Mark Carney and UNFCCC executive secretary, Patricia Espinosa.

The proposed policy solutions were high level: "Pricing externalities, promoting innovation, ensuring financial stability, and ensuring policy coherence."

The more detailed policy prescriptions are set out in the hundreds of other publications, available from an online library.

The final report claims that the Inquiry helped to "shape the shift" in understanding from a resilient financial system, to one aligned with wider environmental and social goals, citing work from around the world that the Inquiry supported or led.

Its legacy was the at-the-time acceleration of attention to sustainability and in particular, systems thinking, within the finance industry.

"I think we helped achieve a recognition that to deliver sustainability outcomes we're going to need a transformation of how the financial system works" Nick Robins said to me.

"We weren't the only people working on this and the UNEP Inquiry was very much a product of its time."

"A key focus was the run-up to the 2015 Paris Agreement, which for the first time included a specific financial goal ['to make financial flows consistent with low-carbon development', Article 2.1(c)]."

"The issue was no longer about this green asset or this green fund. The task was to shift the entire system on its axis – and recognise that as well as market and policy failures in the so-called real-economy, there were real policy and market failures in the financial system too."

"The surge in green and sustainable financial regulation since then has underscored this point, but the vast bulk of financial rules are still blind to their sustainability implications."

"The transformation of the financial system is now embedded into government agreements (such as the COP27 climate conclusions). Eight years on from the Inquiry, we need a new system reform agenda, celebrating how much has been achieved, but also focusing on the areas that still need further work, not least flows of finance into the Global South and the relationship between the financial system, inequality, climate and nature."

"One of the striking things about the Inquiry is we tried to make sure we broke out of the OECD stranglehold on discussions about sustainable finance. We worked in Bangladesh and Brazil, China and India as well as the EU and the US. One particularly important outcome was working with China, which announced at the launch of the Inquiry report much to our surprise that it would be making green finance a priority of its G20 in 2016: this really moved the agenda from a narrowly OECD to a fully global footing."

IIGCC

For its size, IIGCC punches above its weight. Its London offices are not grand. IIGCC does not host big conferences, nor does it require reporting, nor critically assess the actions of its members. Its theory of change is working groups and guides, and more recently, drop-in "surgeries".

This tends to make it a less threatening group than its peers. If the investor is unsupportive, then the guide can be ignored. IIGCC works in the US with Ceres, in Asia with AIGCC (Asia Investor Group on Climate Change), and in Australia and New Zealand with IGCC (Investor Group on Climate Change).

IIGCC tends to be more consensus-driven than PRI, in part, because IIGCC is specific to one topic—climate change, one constituency—investors, and one region—Europe. It did catch the attention of some of its US members when IIGCC publicly pushed back against Europe's plans to include gas in the EU Taxonomy.

IIGCC is organised in three parts, investment practice, corporate disclosure and stewardship, and policy engagement. Each part works through working or advisory groups of members. Members can join, put themselves forward to chair or co-chair and contribute their expertise in the formulation of guides, round tables or policy positions.

IIGCC is one of the groups behind Climate Action 100+, the Net Zero Asset Managers Initiative and the Paris Aligned Investment Initiative (PAII).

The PAII is my go to initiative to understand GHG emissions reporting, scenarios and target setting, how to prepare an effective TCFD report, and how to set and implement a Paris-aligned net zero target.

The PAII was established in May 2019 in response to the growth in net zero commitments. Its centrepiece is the net zero investment framework, which is described "as a practical blueprint enabling investors to decarbonise investment portfolios and increase investment in climate solutions, in a way that is consistent with and contributes to a 1.5°C net zero emissions future" (IIGCC 2021).

I'd agree. It very clearly establishes best practices on emissions disclosures, target setting, approach to both emissions-intensity metrics and climate solutions metrics, and how progress can be made consistent with the Paris Climate Agreement.

It is an impressive piece of work and core reading for responsible investment professionals.

IIGCC's events, industry guides and investment working groups, as well as its policy engagement and policy briefings, are an important contribution to how we understand responsible investment, particularly on climate change.

I asked Stephanie Pfeifer, CEO at IIGCC, about the origins of IIGCC, how it came about, who set it up and why.

Pfeifer said, "IIGCC was established in 2001 on the initiative of the responsible investment leads at Universities Superannuation Scheme (USS) as a forum for collaboration between pension funds and asset managers on issues related to climate change."

"Specifically, USS's beneficiaries – university lecturers – had begun to question the risks that climate change might pose to their pensions. So, it is them that we must thank for initiating the IIGCC journey."

"By the time I joined in 2005, IIGCC still only consisted of a small group of about 20 pension funds and asset managers, representing around £1 trillion in assets."

"Much of our work at this time and throughout the 2000s focused on education, and to a smaller extent on policy engagement and asking companies for disclosure on climate risk. In particular, trustee training explaining the potential physical, policy and reputational risks from climate change but also the potential opportunities from the energy transition. This was still a time when I went to meetings where I was still confronted by climate sceptics – something that has now undoubtedly changed."

"More broadly, at this time there was absent or poor policy and regulation."

"Overall, looking back the journey to date can be characterised as from 'nowhere to mainstream'. Industry progress – which is mirrored by IIGCC's growth so that today we have over 400 members representing over €65 trillion in assets under management – must therefore be contextualised by how far we have come."

GIIN

The Global Impact Investing Network (GIIN) describes itself as "the global champion of impact investing".

GIIN's legal setup is a US nonprofit. Part-funded through membership, part philanthropy, its theory of change is similar to that of other initiatives. It provides tools and resources, industry research and market leadership initiatives.

GIIN established four characteristics of impact investing (GIIN 2019). The wording is a bit clunky, but the characteristics (or a close variation) have been widely accepted as an industry standard. They are:

- Intentionality: An intentional desire to contribute to measurable social or environmental benefit.
- Impact data in investment design: In other words, the intentional desire should be underpinned with data.
- Manage impact performance: Investments should be managed towards that intention. More recently the word used to describe this is "additionality".
- Contribute to the growth of the industry: I interpret this as to report.

GIIN also considers financial risk reward. Today, most institutional investors would consider impact investment to have market rates of return. Investments undertaken by high net worth individuals, which are subject to different regulatory structures, are perhaps willing to tolerate lower rates of return for higher degrees of impact. Institutional investors, bound by fiduciary duties, would not.

One of GIIN's tools is IRIS+, which provides investors with indicators to track impact. IRIS+ has established a series of metrics which investors can use across themes to measure—and make progress towards—impact goals.

IRIS+ helps impact investors understand themes, indicators per theme, and how this links to the SDGs.

The work GIIN does is useful, its staff are senior and competent. Investors that use GIIN publications are quick to explain their usefulness. But I wonder if GIIN has been held back by its US headquarters, and the background of populist US politics and the politicisation of impact topics.

Another possibility is that impact investment is a more distinct activity in the US (and Switzerland). The US has a higher concentration of high net worth individuals, in a lower tax jurisdiction, with distinct legal vehicles for impact investment, and therefore, where high net worth individuals actively seek impact investment products.

In Europe, it is the Impact Management Project (IMP) that many investors refer to, and to which GIIN contributed.

The IMP sets out five dimensions of impact framework:

- What: What intended social and/or environmental outcomes occur? How are these outcomes measured and reported on?
- Who: Who is experiencing the outcomes?

- How much: What is the scale of the outcome (size) and what is the depth of the outcome, and whether the project or company is likely to have been otherwise funded?
- Contribution: What is the contribution (attribution) of the enterprise to the outcome?
- Impact risks: What are the potential adverse impacts and sustainability risks involved and how are they managed?

And because impact is often place-specific and requires narrative-based reporting, impact reporting tends to be principles-based.

A question for both frameworks is "additionality"—how the investor considers their contribution to impact.

When it comes to impact, typically an investor would think about a theme where the investor has expertise, where the investor can make an impact, and potential client base.

It was US Supreme Court Justice, Potter Stewart who is remembered for his non-definition of obscenity: "I know it when I see it." Perhaps the same could be said for impact.

It's these types of questions that determine the need for a group like GIIN.

ShareAction

Headquartered in the UK, ShareAction is a credible campaigner. ShareAction has managed to retain the fine line of credibility with pension funds and asset managers, but also accountability, not straying from commitments made to a range of sustainability topics, even if not universally popular.

ShareAction tends to be a few years ahead of the conventional interpretation of responsible investment. In other words, they're a good group to track.

Their activities tend to be a mix of events, research, guides and scorecards, ranking pension funds or asset managers on their approach to sustainability topics. ShareAction even publishes draft legislation, often with the help of pro bono lawyers.

"ShareAction's origins are a student-led campaign in the university sector, originally targeting USS, and the group was called Ethics for USS", CEO Catherine Howarth told me.

"This was in the late 1990s. It started with students but quite quickly and effectively started to get members of the university scheme on board."

"USS was the largest scheme in the UK, with an informed membership, uniquely well-educated by definition, about responsible and ethical investment."

"As a result of the campaigning, USS became the first scheme in the country to employ responsible investment officers and has been active in setting up global initiatives from PRI to CA100+ [and, IIGCC]."

"It was a scheme-specific campaign to begin with. A group of civil society organisations, including Oxfam, WWF and trade unions, then chose to sponsor into existence a new NGO building on what had happened at USS and expanding the campaign for responsible investment to the wider pensions and investment industry."

"The NGO was founded in 2005, it became a charity in 2006, and I joined in 2008."

ShareAction does not receive membership funding from investors (unlike PRI, IIGCC or UKSIF) and so tends to take a stronger line.

It does however mean that their work programme can, at times, be determined by a complex mix of ShareAction's strategy, trending topics and whether it sparks a donor's interest.

ShareAction does undertake engagement with individual savers, although my sense is that it is done with mixed success. In some respects, though, it doesn't matter. That ShareAction has the capability to do so, is enough motivation for pension funds to respond to ShareAction's activities.

ShareAction also coordinates resolutions that are filed at company AGMs. In 2022, ShareAction convened a group of 10 institutional investors to file at UK supermarket Sainsbury's AGM calling on Sainsbury's to pay its staff a living wage (ShareAction 2022).

ShareAction was often asked why Sainsbury's. "Because," staff would reply, "for Sainsbury's, a living wage is within reach, that's not the case for some other supermarkets." Other supermarkets were of course watching closely.

Sainsbury's committed to pay a living wage to its directly employed staff, but not its contractors, and did not commit to sign up to the Living Wage Foundation (which would require Sainsbury's to continue to raise wage levels in line with the Foundation's living wage calculations).

The co-filers, which included the UK's Brunel Pension Partnership, NEST and LGIM, decided not to withdraw the resolution, and it went on to receive 17% of the shareholder vote. Not a majority, but enough to keep the pressure on Sainsbury's.

Asset managers that did not support, such as Schroders, were asked to explain their approach in the responsible investment trade press (Responsible Investor 2022).

NZAMI

To great fanfare, Mark Carney launched GFANZ, the Global Financial Alliance for Net Zero, at COP 26. GFANZ is an umbrella group of net zero groups, including the Net-Zero Banking Alliance (NZBA), Net Zero Asset Managers Initiative (NZAMI), Net-Zero Insurance Alliance (NZIA), the Net Zero Asset Owners Alliance (AOA) and the Net Zero Investment Consultants Initiative (NZICI).

Some groups are initiatives, some alliances, some groups hyphenate net-zero, some do not, and in turn, each initiative is supported by a range of investor groups. It's complicated.

The initiatives published the AUMs of the organisations that have signed up, even if the organisations had only a partial net zero target. Indeed, many of the asset managers signed up to NZAMI have set targets that are something like, "by 2030, 90% of the companies in which we invest will be net zero aligned." This feels like a long lead time for companies to be net zero aligned.

This then requires another initiative, the Science-Based Targets initiative (SBTi) to determine whether the company is net zero aligned. Net zero aligned does not mean the company is net zero, it means the company is taking steps to become net zero, and the pathway the company is taking is consistent with the Paris Climate Agreement.

SBTi is staffed mostly through staff secondments from other sustainability initiatives. Companies must pay to receive SBTi "certification".

Disclosure of a company's decarbonisation strategy, subject to third-party scrutiny, is nevertheless an important step forward.

In September 2022, two pension funds, including Australia's Cbus, known for being a sustainability leader, left the AOA, both citing resourcing issues (paradoxically, there's a fee, and a relatively considerable one, to join the AOA, but not the NZAMI). Vanguard left the NZAMI in late 2022. Swiss Re left the Net-Zero Insurance Alliance, joining Zurich, Hannover Re and Munich Re. It's reported that a number of banks are considering following suit and leaving the NZBA.

Until around 2021, commitments were aspirational. To their credit, the sustainability groups have strengthened reporting requirements, and now the lawyers and politicians are involved, and they're pushing back.

If a bank or investor is telling its clients to buy products and uses commitments to net zero in its marketing materials or even in a generic form on its website that it is not able to, or not willing to meet, then this is greenwashing, and potentially mis-selling, which may explain why in-house legal teams are advising their firm to leave the initiative. There are also issues around acting in concert, where firms are perceived to have colluded.

Another problem is that few—even within the sustainability community—believe that limiting warming to 1.5 degrees Centigrade is possible. It's like implementing a 20kph speed limit in the hope that cars go 25kph. It shouldn't be like that of course, but it often is.

This could be perceived as a fiduciary risk for an investor that commits to net zero GHG emissions by 2050 so as to limit warming to 1.5 degrees Centigrade, meanwhile expecting net zero GHG emissions not in 2050, but a later date. This is a challenge many investors and banks are now working through, and it's triggering some to renege their signatory.

The NZAMI was spearheaded by Generation Investment Management.

My guess at the time was that Generation realised that, we were not going to get close to net zero without the big asset managers involved, and that, of the many mid-sized asset managers committed to sustainability, Generation was a natural convenor.

I asked David Blood why Generation was one of the firms that started NZAMI. "In July 2020, Generation committed to align our clients' investment portfolios with net zero emissions by 2040."

"We made this commitment to our clients because leadership on the climate crisis is critical, and we believe that managing climate risk and opportunity is inseparable from our fiduciary responsibility. In the months following our commitment, we worked with peers and partners, in particular the Institutional Investors Group on Climate Change (IIGCC), to establish a new Net Zero Asset Managers Initiative (NZAMI) – a coalition of like-minded managers committed to investing in line with net zero emissions by mid-century."

The NZAMI had waves of sign ups, including the firm I worked for. My reading was that it was inevitable that all asset managers would have to sign at some point (a similar dynamic to PRI), and that, as an initiative, it was one worth supporting, with serious leadership from Generation and others.

The first requirement of the NZAMI is to set a target and to disclose the assets under management to which the target applies. The second requirement (which follows one year later) is to report on progress against that target. In 2022, reporting was via the Carbon Disclosure Project (CDP), given the issues PRI had with its reporting framework. In 2023, via CDP or PRI.

For European investors, the NZAMI is supported by IIGCC who do a good job of providing guides, drop-in surgeries, workshops and so forth to support signatories in their disclosure.

Despite cracks appearing in the GFANZ initiatives, net zero target setting, measurement and disclosure are important steps forward. If capital allocation is to be a lever of influence, it needs to be at scale and that requires collaboration.

An analogy I use is town centre parking. A council wants to encourage its residents to take the bus, rather than drive. So the council introduces parking restrictions in the town centre. Residents park in nearby streets. So the council introduces parking restrictions in nearby streets. And so forth. Until, the drive is not convenient, and residents take the bus instead.

To apply to responsible investment, here, an investor wants to encourage a company to decarbonise, so it divests. Another investor follows, and another, and so forth. Until, the company is no longer profitable, and either decarbonises, or returns capital to shareholders, or the company is subject to private or state ownership (and unwinds).

I think NZAMI will have some real-world impact, but only if target setting is meaningful and investors at scale divest from companies, or consistently vote against directors from companies, that fail to transition or have inadequate transition plans.

In turn, it will give companies more confidence to decarbonise their products at cost and invest in environmental solutions.

The risk is that there remains enough investors outside the initiative (and indeed, inside the initiative) that are comfortable investing in high carbon revenue-generating companies often at cheap prices. The next few years will determine whether net zero capital allocation is an effective form of influence, rather than (or perhaps, as well as) collaborative stewardship, such as Climate Action 100+.

Finally, perhaps one thought for the stakeholder groups: I think there is an issue with the way the net zero groups influence responsible investment more generally. Climate change has very specific characteristics. It is really quite unlike many other sustainability issues in a number of ways.

Potentially, there is an element of the tail-wagging-the-dog as responsible investment seems to see itself in the image of net zero initiatives. Rather than, responsible investment having a clear idea of its own dynamics and seeing how the approaches used for other issues fit into climate change.

Sofia Bartholdy, Net Zero Lead at the Church Commissioners for England, told me, "The incredibly quick uptake in net zero targets is a very positive development, but with the first targets being set in 2019, no one has experience in implementing these targets."

"This is true for investors as well as governments, companies and other non-state actors."

"Collaborative efforts such as AOA and NZAMI provide forums to speed up the learning curve among investors."

"I find the level of collaboration and the willingness to learn from each other a strength of the responsible investment industry."

Make My Money Matter

Make My Money Matter is a UK campaign group. There are similar groups in most major markets. MMMM was co-founded by screenwriter and film producer Richard Curtis (famous for Four Weddings and a Funeral and Love Actually). British businesswoman (and BBC Two television show, Dragons' Den) Deborah Meadan is a vocal supporter.

On MMMM's part, there appears to be little love for the UK pensions industry and more recently, the UK banking industry. This is due to MMMM's view that the pensions industry has been too slow to respond to and address the seriousness and urgency of climate change.

MMMM has coined the term, the "21 times challenge". "21x: It's the most powerful thing you can do to protect the planet. Cut your carbon 21 times more than going veggie, giving up flying and switching energy provider simply by making pension green."

MMMM tells its supporters that "by encouraging your [pension] provider to clean up their investments, you can cut your carbon footprint … from the comfort of your sofa." "That means you can ensure the £3 trillion in UK pensions flows to business that are trying to tackle climate change".

But MMMM is campaigning for pension funds to put in place a net zero target, even for pension funds that are de-risking (not accepting new members and working towards buy-out).

A net zero target does not mean that the pension fund is investing in businesses that are "trying to tackle climate change", as MMMM's website says. A pension fund may make an allocation to a green solutions fund, but that is certainly not indicative of the £3 trillion UK pension market, nor is it what MMMM is currently campaigning for.

(Not least, because for many pension funds, a large part of their portfolio will be allocations to government bonds, and in the UK's case, allocations to UK government bonds.)

Through neat info-graphics, social media and events, MMMM takes their message to the public, and then encourages the public to, in turn, contact their provider. In practice few do. It is a niche action. But it has caught the attention of the pension funds, their trustees and staff. MMMM names and shames pension funds that the group believes are failing to take appropriate levels of action.

I'm in two minds. On the one hand, MMMM is trying to connect a saver's pension fund with the real-world impact of the pension's investment decisions. This is a good thing. We tried to do this at Oxfam. MMMM has been considerably more successful. It is inspiring a new group of campaigners to understand pensions and to consider how pensions can support or hinder environmental goals.

But the ask requires further thought. This is where substance and social media come into conflict. Calling on pension schemes to use their stewardship rights to co-file, vote against directors, allocate to sustainability solution funds, and a series of other actions that will deliver real-world sustainability impact, is less catchy than, "make a net zero target".

But I'm sure MMMM could find a way to strengthen their ask and keep it catchy. MMMM is evolving (and I believe, improving) its approach, but on my part, I felt MMMM's early campaigns were not well researched, perhaps even, a bit lazy.

Nor, and I've given this quite a bit of thought, does a "green" pension fund abscond an individual from the set of individual actions needed to lower our carbon footprints, eating less meat, flying less and changing energy providers, that the 21X challenge implies. In fact, I think this is counterproductive.

I wish MMMM success, but I hope it changes its approach. Whatever the pensions industry or banking industry does, MMMM will say it's not enough, with the campaigns becoming less about actual change, and more about social media 'likes'. Such an organisation has its place. But the ask needs more substance. Failure to do so could disenfranchise a group of (mostly young) campaigners essentially campaigning for the wrong thing.

References

BCG (2021), The Biodiversity Crisis Is a Business Crisis. [online]. Available from: https://www.bcg.com/publications/2021/biodiversity-loss-business-implic ations-responses (Accessed, January 2023).

IIGCC (2021), Net Zero Investment Framework Implementation Guide. [online]. Available from: https://www.iigcc.org/resource/net-zero-investment-framework-implementation-guide/ (Accessed, January 2023).

GIIN (2019), Core Characteristics of Impact Investing. [online]. Available from: https://thegiin.org/characteristics/ (Accessed, January 2023).

Responsible Investor (2022), Schroders asked to leave fair pay group after Sainsbury's living wage vote. [online]. Available from: https://www.responsible-inv

estor.com/schroders-asked-to-leave-fair-pay-group-over-sainsburys-living-wage-vote/ (Accessed, January 2023).

ShareAction (2022), Sainsbury's Living Wage resolution achieves significant shareholder support. [online]. Available from: https://shareaction.org/news/sainsburys-living-wage-resolution-achieves-significant-shareholder-support (Accessed, January 2023).

Washington Examiner (2022), JPMorgan CEO snaps back Rep Tlaib fossil fuels. [online]. Available from: https://www.washingtonexaminer.com/restoring-america/faith-freedom-self-reliance/jp-morgan-ceo-snaps-back-rep-tlaib-fossil-fuels (Accessed, January 2023).

8

The Detractors

Don't get Fancy

In November 2021, Prince Harry and Meghan, Duchess of Sussex backed an impact fund (New York Times 2021). In April 2022, Brazilian fashion model, Gisele Bundchen, called herself an "ESG advisor" (iGamingFuture 2021).

Responsible investment was in vogue.

But, alongside new-found cheerleaders were a growing number of critics of responsible investment.

Let's first start with two individuals, who while smart and eloquent, had only brief flirtations with responsible investment, neither long enough to be, in my view, credible. Still, their criticisms are widely reported, and I'll turn first to Tariq Fancy.

Tariq Fancy is the former Ex-CIO for sustainable investing at Blackrock, the world's largest asset manager. He was there for just a couple of years. In August 2021, Fancy published the first in a trilogy of essays titled, "The Secret Diary of a 'Sustainable Investor'" (Fancy 2021). He's since made a name for himself as a vocal critic of responsible investment.

His essay starts at a Swiss investment conference.

"Asked a thoughtful question by a Swiss client on how these investment vehicles actually contribute to fighting climate change, I explained how high growth of these products might, in theory, find some way to indirectly increase financing costs for higher carbon-emitting companies, incentivizing them to lower emissions."

© The Author(s), under exclusive license to Springer Nature
Switzerland AG 2023
W. Martindale, *Responsible Investment*, https://doi.org/10.1007/978-3-031-44536-1_8

Fancy was reprimanded by colleagues for failing to speak to a set of pre-agreed talking points. "Keep it simple." "All they need to know is that it has a lower carbon footprint — they should do it to fight climate change."

Fancy says he "sincerely believed" that responsible investment was a "step in the right direction", but that he now realises that he was wrong.

His second essay argues that motivation for corporate sustainability must be rooted in shareholder value (citing Blackrock's second operating principle: "we're passionate about performance"). As such, there is futility in shareholder engagement. Fancy says that there is danger in leaving issues as serious as the climate crisis to voluntary action, rather, the answer is to change the rules of the game and in the case of climate change that would mean a meaningful price on carbon.

The third essay doesn't hold back. It calls sustainable investment "a dangerous placebo that harms the public interest" and "like selling wheatgrass to a cancer patient." He goes on, "this is greenwashing our entire economic system."

He says, "the more I explained the concept of sustainable investing to people, the more they seemed eager to believe that it would help — and slightly relieved that it might allow us to leave things the way they are."

In polling commissioned by Fancy, he found that, "every time people read the latest such headline about guarding against climate change-related risks in the financial system, they mistakenly believed that these efforts were helpful in the fight against climate change itself."

The distraction, Fancy argues, is real.

In short, responsible investment has no real-world impact, that capital allocation is irrelevant and that voluntary action is insufficient, that worse, sustainable investment gives the perception of real-world impact, slowing the policy changes that are necessary.

His essays end with:

> "It's time to accept that there are nine words that we need to hear, because we can only build a sustainable future once we're no longer terrified to hear them. I'm from the government and I'm here to help."

In December 2021, US climate enjoy John Kerry said "I believe the private sector has the ability to win this [climate change] battle for us" (CNBC 2021). Other policymakers have made similar comments. Such remarks appear to play into Fancy's argument that policymakers believe it is private capital–not public policy change–that will help us achieve our sustainability goals.

Speaking at a conference in London, academic and sustainable finance expert, Alex Edmans, was quick to take apart Fancy's arguments (Net Zero Investor 2022).

Edmans said, "There are two broad claims made by the ESG industry. First, it leads to financial returns. Second, it has real-world impact in terms of social returns." "Both of those claims are partially true. Some things work and some things don't."

Edmans added, "Why are we even responding to Tariq Fancy to begin with? He was at one asset management firm for just two years. He's incentivised to only have a negative view."

The second is Stuart Kirk.

Stuart Kirk shot to responsible investment fame following a witty but careless speech at a conference in London.

Kirk was, for a year, global head of responsible investment at HSBC asset management. Clearly unafraid of controversy, Kirk said, "who cares if Miami is six feet underwater." His LinkedIn profile was later updated to say, "actually loves Miami" while he works on a new project.

His speech was at a FT Moral Money summit in May 2022 (FT 2022a). Kirk told investors, they "need not worry about climate risk."

Referring to a previous speaker, "[she] said we're not going to survive, and indeed no-one ran from the room, in fact, most of you barely looked up from your mobile phones." "It's become so hyperbolic," he said standing in front of a slide titled, "Unsubstantiated, shrill, partisan, self-serving, apocalyptic warnings are always wrong."

A follow-up article titled "ESG must be split in two" published a few months later is of more interest. If there was a plan here, notoriety followed by substance, then kudos. But I suspect not. Kirk would have known his speech would attract attention. But there are hints of contrition in the follow-up article (FT 2022b).

"I'm pro-ESG, as it happens," he says. "But I have long argued it has an existential defect … The flaw is that ESG has carried two meanings from birth. Regulators have never bothered disentangling them, so the whole industry speaks and behaves at cross purposes."

"Two completely different meanings then. One considers E, S and G as inputs into an investment process, the other as outputs - or goals - to maximise. This conflict leads to myriad misunderstandings."

Here, Kirk is right (although, many others had already made this distinction). Indeed, the Fiduciary Duty in the 21st Century US Roadmap made this clear in 2016, as did the European Commission's Action Plan on Sustainable Finance in 2018.

"In an ESG-input world," Kirk said, "it is OK to own a polluting Japanese manufacturer with terrible governance if these risks are considered less material than other drivers of returns."

"Fund reporting," he goes on, "is a nonsense when ESG has two meanings." His recommendation is to "disentangle the two."

The Economist

In late July 2022, the Economist published a leader and special supplement, titled "ESG should be boiled down to one simple measure: emissions. Three letters that won't save the planet" (The Economist 2022a).

I found the article's conclusion somewhat petty: ESG is "an abbreviation that is in danger of standing for exaggerated, superficial guff."

The article argues that ESG suffers from three fundamental problems:

"First, because it lumps together a dizzying array of objectives, it provides no coherent guide for investors and firms to make the trade-offs that are inevitable in any society."

"The industry's second problem is that it is not being straight about incentives … it is often very profitable for a business to externalise costs, such as pollution, onto society rather than bear them directly. As a result the link between virtue and financial outperformance is suspect."

"Finally, ESG has a measurement problem: the various scoring systems have gaping inconsistencies and are easily gamed."

Instead, "it is better to simply focus on the e," "the e should stand not for environmental factors but for emissions alone."

There are at least three arguments against.

First, ESG issues interconnect. Biodiversity, nature, water and a range of environmental issues are the other side of the emissions coin. But so too are social issues. It is the poorest who are most vulnerable to climate change and the richest most likely to cause it.

Second, a broad range of ESG issues is necessary to understand a company's financial prospects but also to understand a company's contribution to solutions. We can't get that from just emissions.

Third, emissions metrics themselves require scrutiny, which the article does not address. Emissions metrics depend on the type of industry, mix of energy sources, complexity of supply chain and so forth, data that can be found from a broader set of ESG metrics. Portfolio design that solves to a headline emissions target is unlikely to address rising emissions.

In other words, the same criticisms the Economist levels at "ESG" can also be levelled at just emissions. Scope 3 emissions disclosures, necessary for company comparisons, are often incomplete or inaccurate.

The Economist's analysis on S issues were interesting. "In a dynamic, decentralised economy individual firms will make different decisions about their social conduct." "There is no one template."

Because S issues are company, sector and country-specific, an attempt to quantify a range of S issues into a single score is not decision-useful. Here, I would agree.

But in the same way that investors would have different expectations of companies operating in high-carbon industries, so too would investors have different expectations of companies operating in countries with less well-established social foundations. Paying a living wage in the UK is not the same as paying a living wage in, say, Bangladesh where there are many more instances of underpayment. Black-box, subjective scores on S (and E) issues are not particularly helpful. But that does not mean they should be ignored, rather the investor should consider how to access and assess source data.

The rebuttals to the article were not well coordinated, and in some respects, played into The Economist's central argument that ESG has "morphed into shorthand for hype and controversy."

Sir Ronald Cohen's response was off-the-mark. "I estimate that more than $2TN of global capital is invested according to impact principles," he said. He goes on to speak about the ISSB, which is not an impact framework. "Big data, artificial intelligence and machine learning are enabling us to measure and value physical impacts and integrate them into financial analysis and the valuation of companies," he says (The Economist 2022b).

Fiona Reynolds, former CEO at PRI said, "it's really quite simple," "ESG principles can all be boiled down to a few simple truths: we all have responsibilities as citizens, and we should all act ethically, and we should invest in ways that don't contribute to human suffering or to the detriment of the planet we all inhabit" (Top1000 Funds 2022).

My view is that it is not quite that simple, because ESG is primarily a risk term, not as Reynolds sets out, an ethical one.

Nathan Fabian, Chief Responsible Investment Officer at PRI said, "it's vital that we tackle emissions, but investors and policymakers alike cannot treat the issue in isolation. A holistic approach is needed" (The Economist 2022b).

Richard Roberts, Volans' Inquiry Lead said—and I agree with every word—"the problem isn't ESG as such, but rather the obsession with being able to boil performance down to a single number. If the length you're willing to go to 'integrate' ESG is to look at a single column in a spreadsheet, then yes, it's probably better to have the column be something simple and comparable like GHG emissions, rather than some opaquely constructed ESG score. But you have to have a pretty low opinion of the financial industry to think that's really the best they can do."

Finally, Hugh Whelan's, writing in Responsible Investor, provided another robust response. "If the Economist does a critical 'special' and Musk launches a Twitter missile, we're making headway!" "Finance has power and responsibility in a market economy." "How it responds is paramount. It is not just about the impact on investment returns, but the materiality to society" (Responsible Investor 2022).

Yale School of Management

Published in March 2023, academics Hartzmark and Shue used a percentile-based approach to classify firms as green, neutral, or brown, in their paper, "Counterproductive Sustainable Investing: The Impact Elasticity of Brown and Green Firms"

Kelly Shue is a professor at Yale School of Management, Samuel Hartzmark is a professor at the Carroll School of Management at Boston College.

The authors first calculated each firm's emissions intensity, which is defined as Scope 1 and Scope 2 emissions scaled by revenue. They then ranked firms by their emissions intensity and divided them into quintiles, with the lowest quintile representing the greenest firms and the highest quintile representing the brownest firms. Firms in the middle three quintiles were classified as neutral.

This means that service-based companies were categorised as "green", while energy, transport and agricultural companies were categorised as "brown". It also means that an oil and gas company that is transitioning, consistent with the Paris Climate Agreement, may still be considered a brown firm due to its overall emissions intensity.

It also favours companies that have outsourced their higher carbon industrial activities (and as such, have high Scope 3, butrelatively low Scope 1 and Scope 2 emissions).

Here are the paper's findings:

- Brown firms have a greater negative impact elasticity than green firms. This means that increasing the cost of capital for brown firms leads to larger negative changes in their environmental performance, while reducing the cost of capital forgreen firms leads to smaller positive changes in their environmental performance.

- Sustainable investing that directs capital away from brown firms and towards green firms may be counterproductive. This is because it may make brown firms more brown without making green firms more green.

The paper's abstract says:

"We show empirically that a reduction in financing costs for firms that are already green leads to small improvements in impact at best ... the sustainable investing movement primarily rewards green firms for economically trivial reductions in their already low levels of emissions."

A follow-up podcast on the popular channel, "Freakonomics", featured the paper's author, asking "Are ESG investors actually helping the environment?", concluding that, no, they are "probably not".

There are a range of issues with the paper

First, it's a gross simplification to categorise the "sustainable investing movement" as just focused on GHG emissions.

Second, it's another gross simplification to categorise "green" as top 20% of companies' Scope 1 and 2 GHG emissions by sales and "brown" as bottom 20% of companies' Scope 1 and 2 GHG emissions by sales.

Third, it's incorrect to infer that this is the "dominant" approach of 35 trillion USDs of invested capital.

Finally, it's therefore not possible, based on the research, to conclude, as the podcast did, that ESG investors are not helping the environment.

As with The Economist article, there were several rebuttals.

But when authoritative organisations, such as The Economist, Freakonomics and Yale School of Management are misinterpreting responsible investment, it is perhaps worth asking ourselves if the responsible investment industry has lost control of its own narrative.

Republican Law-Makers

In recent months, vocal criticism of responsible investment has come from US populists, invoking culture wars. One example (of many) is Ron Paul.

Ron Paul is a libertarian. He was a Representative from Texas off-and-on from 1976 through to 2013, running for president in 2008 and 2012 for the Republican party.

He's in his late 80s. He's a prolific tweeter, although whether or not it's him actually tweeting is to be debated.

In October 2022, he tweeted, "More and more people are waking up to the absurdity of "getting rid of fossil fuels" and "net zero" carbon. The idea of "eating bugs" is dehumanising. We work to bring home the bacon, not the bugs. "Woke capitalism" & "climate finance" are massive bubbles waiting to burst" (Paul 2022).

For some months, "ESG" has been under attack from US Republican lawmakers.

Some States have introduced, or talk of introducing, regulations that would prohibit what they consider to be ESG-related activities. In other states it is limited to public questioning of investors from treasurers, attorney generals or elected representatives, questions investors would no doubt prefer to avoid.

In 2022, a group named "State Financial Officers Foundation" published a video titled "Our Money Your Values - What is ESG?" (Our Money Your Values 2022).

It starts with a calm voice-over asking, "Have you heard about this thing called ESG?" "On its surface, ESG claims to have a well-intended objective. To promote corporate responsibility." It switches to a low man's voice. "But under the surface, lurking inside the ambiguous language … is something very different. ESG policies are a back-door that progressives are using to invade our economy so they can advance their radical economic, social and climate agenda."

"ESG undermines democracy. ESG can't win in the courts. It is infiltrating corporate America, with the help of unelected bureaucrats. Like the shadow government. With no accountability. Attack our energy independence. Small business. Family farms."

"They even want to ban wheatfarming." "And to mandate abortions on demand."

It goes on.

It is straight from the Republican-right playbook.

In large part, the attacks on ESG issues are because of the perception, particularly, but not uniquely in the US, about what is the appropriate role of government versus the private sector.

The notion in the US is often that when there are large-scale problems it is the role of the government (not markets) to fix it.

The role of investors is to make money. It is the role of the government to solve climate change (or, for State Financial Officers Foundation, not to fix climate change).

But, even if that is true, it doesn't take away from the financial materiality of these issues. The issues affect the market, even if, as some argue, it is not for investors to affect the issues.

Academic and Forbes columnist, Bob Eccles said in a post:

"In contrast, and to show you how different our views are [to the anti-ESG Republicans], I see ESG as an idea which is actually strengthening our free markets through better capital allocation. ESG is simply about the sensible management of material risk factors by both companies and investors." (Forbes, 2022)

In an interview with me, Bob Eccles stressed the importance of financial materiality and investment choice:

"It's bullshit to say, don't take account of these things [ESG issues]. That's a political statement to say don't do ESG integration."

"[With regard to impact] you need to distinguish between whether impact is concessionary or non-concessionary, and whether it's your money or someone else's."

"If you want to invest in a MAGA fund, if you want to invest according to your republican values, and it's your money, go ahead."

"[But] if it's an impact investment, and it has the same risk adjusted returns, you're not violating your fiduciary duty."

I expect the SFOF (and others) will run with this for some time to come. And their videos, while farcical in substance, are well done in style. And by conflating ESG processes and products, our terminology is confusing. It's easy for SFOF to find a way through, spurred on by other vocal anti-ESG critics, such as Elon Musk, who in January 2023 tweeted that "The S in ESG stands for Satanic" (Musk 2023).

I welcome the critique. It demonstrates that ESG matters and it will force us to up our game. Regardless of Republican talking points, there are problems in society that cannot be fixed without the conscious participation of the private sector.

I asked Fiona Reynolds about the recent Republican push-back against responsible investment. "On the one hand, I can't see responsible investment going backwards. It is now the major lens for how we consider capital allocation and investment decision-making."

"But at the same time, I do think that we need real acceleration in how we think about real-world sustainability impact and we've got this problem in the United States of criticisms of woke-capitalism and it is slowing things down. Capital is global and the US is the biggest financial market in the world."

"We were starting to move to a point where we can get global definitions and standards. If the Republicans win the next presidential election, it could be catastrophic for responsible investment in general."

Reynolds believes it's important that investors engage with lobbying.

"One of the problems is that, as investors, we've not adequately dealt with the issue of political lobbying, the people funding the politicians. We take a knife to a gun fight. We're not tactical. I don't believe that there is any organization in the responsible investment space doing the big overarching political strategy piece."

Richard Roberts, Inquiry Lead at Volans, said, "There's a blowback against ESG and the responsible investment groups are asking, 'what are we going to do to defend it?' But it's not their job. If you as an investor believe ESG integration is adding value you should defend it. You shouldn't expect PRI to come to your defence."

David Blood said, "We are unsurprised by the recent backlash against the multiple definitions and confusing terminology, the over-reliance on checklists, the potentially misleading marketing campaigns, and the frequent lack of rigor and accountability."

"But these criticisms are by no means evidence that sustainable investing and ESG are failed concepts. Instead, they are welcome challenges to ensure that sustainable investing and the incorporation of ESG factors are carefully defined, clearly understood and effectively practiced."

"Sustainable investing is consistent with the fiduciary duty that investment professionals owe their clients. Those who don't take sustainability factors into account aren't fulfilling that duty."

"Banning consideration of ESG factors would not only lead to poor investment outcomes; it would constitute a clear dereliction of fiduciary duty."

References

CNBC (2021), John Kerry Says Private Sector Can Win Climate Change Battle. [online]. Available from: https://www.cnbc.com/2021/12/01/john-kerry-says-private-sector-can-win-climate-change-battle.html (Accessed, January 2023).

Sir Ronald Cohen (2023), Sir Ronald's Philosophy. [online]. Available from: https://sirronaldcohen.org/sir-ronalds-philosophy (Accessed, January 2023).

The Economist (2022a), ESG Should Be Boiled Down to One Simple Measure: Emissions. [online]. Available from: https://www.economist.com/leaders/2022/07/21/esg-should-be-boiled-down-to-one-simple-measure-emissions (Accessed, January 2023).

The Economist (2022b), Letters to the Editor. [online]. Available from: https://www.economist.com/letters/2022/08/11/letters-to-the-editor (Accessed, January 2023).

Elon Musk (2023), Twitter. [online]. Available from https://twitter.com/elonmusk/status/1614786284547112961 (Accessed, January 2023).

Fancy (2021), The Secret Diary of a 'Sustainable Investor'—Part 1, 2 and 3. [online]. Available from: https://medium.com/@sosofancy/the-secret-diary-of-a-sustainable-investor-part-1-70b6987fa139 (Accessed, January 2023).

Forbes (2022), My New Year's Resolution To Republican Politicians Regarding ESG, [online]. Available from: https://www.forbes.com/sites/bobeccles/2023/01/01/my-new-years-resolution-to-republican-politicians-regarding-esg/ (Accessed, January 2023).

Forum for the Future (2021), The Future of Sustainability: Looking Back to Go Forward. [online]. Available from: https://youtu.be/YKHTfDTuxq4 (Accessed, January 2023).

FT (2022a), HSBC Banker Draws Fire After Accusing Policymakers of Climate Change Hyperbole. [online]. Available from: https://www.ft.com/content/1716d9fb-4c8f-4d05-a6b4-bb02a48a73a4 (Accessed, January 2023).

FT (2022b), Stuart Kirk: ESG Must Be Split in Two. [online]. Available from: https://www.ft.com/content/4d5ab95e-177e-42d6-a52f-572cdbc2eff2 (Accessed, January 2023).

IGamingFuture (2021), DraftKings Appoints Gisele Bündchen as ESG Advisor. [online]. Available from: https://igamingfuture.com/draftkings-appoints-gisele-bundchen-as-special-advisor-for-esg-initiatives/ (Accessed, January 2023).

Net Zero Investor (2022), Net Zero Investor Annual Conference. [online]. Available from: www.netzeroinvestor.net (Accessed, January 2023).

New York Times (2021), Harry and Meghan Want to Make Sustainable Investing Mainstream. [online]. Available from: https://www.nytimes.com/2021/10/12/business/dealbook/harry-meghan-ethical-investors.html (Accessed, January 2023).

Our Money, Your Values (2022), What is ESG? [online]. Available from: https://ourmoneyourvalues.com/ (Accessed, January 2023).

Responsible Investor (2022), In Defence of ESG: A response to The Economist. [online]. Available from: https://www.responsible-investor.com/in-defence-of-esg-a-response-to-the-economist/ (Accessed, January 2023).

Top1000 Funds (2022), ESG: It's really quite simple. [online]. Available from: https://www.top1000funds.com/2022/08/esg-its-really-quite-simple/ (Accessed, January 2023).

Paul (2022), Ron Paul. [online]. Available from: https://twitter.com/ronpaul/status/1583849055028801536 (Accessed, January 2023).

YouTube (2022), HSBC's Stuart Kirk Tells FT Investors Need Not Worry About Climate Risk. [online]. Available from: https://www.youtube.com/watch?v=bfNamRmje-s (Accessed, January 2023).

9

Climate First

COP 21

While this book is not a book about climate change, climate change does differ from other ESG issues. It is underpinned by science (in a way that, say, inequality is not). It is time-limited. It is clearly financial. And the political direction is clear (again, in a way that inequality is not).

That's why I've focused more on climate change than other ESG issues.

In this chapter, I'll share some of my notes from COPs 21 and 26, some of the key regulations and concepts, including France's Article 173 and TCFD, how in my previous roles I've set about calculating scenarios, metrics and targets and alignment metrics, the rise of net zero, more on the Global Financial Alliance for Net Zero (GFANZ), climate tilts and green bonds, the Inevitable Policy Response to climate change, carbon markets and greenwashing.

COP, which stands for conference of parties, is essentially a UN decision-making group. The parties are countries that commit to a convention, in this case, the United Nations Framework Convention on Climate Change (UNFCCC), originally signed at the Earth Summit in 1992. COP 1 took place in 1995 and so, COP 21 took place in 2015 in Paris.

The 2015 COP achieved a landmark agreement by governments to limit global warming to well below 2 degrees Centigrade, and towards 1.5 degrees Centigrade (UNFCCC 2015). Having made the commitment, governments would have five years to publish nationally determined contributions.

Article 2 of the agreement commits to:

© The Author(s), under exclusive license to Springer Nature
Switzerland AG 2023
W. Martindale, *Responsible Investment*, https://doi.org/10.1007/978-3-031-44536-1_9

- Hold the increase in global average temperature to well below 2 degrees Centigrade above pre-industrial efforts
- Pursue efforts to limit the temperature increase to 1.5 degrees Centigrade
- Increase the ability to adapt to climate change
- Align finance flows with a pathway to low GHG emissions

This would require countries to reach peak emissions and decarbonise. Article 4 speaks to climate justice, requiring developed countries to "take the lead" and that "support shall be provided" to developing countries. Article 10 is the obligatory nod to technology and Article 12 to education.

Veterans of the climate movement were jubilant. "We've done it." The conference was wonderful.

My favourite speech took place outside the main conference venue. Achim Steiner, then head of the UN Environment Programme (UNEP), was speaking at a reception at the "Musée de l'Homme", just across the river from the Eiffel Tower. In a trilingual German, French and English speech, Steiner talked about the role women played in achieving the Paris Climate Agreement, in particular, then head of the UNFCCC, Christiana Figueres. He compared the literal translation of Musée de l'Homme as museum of humankind—rather than museum of man. He—and Figueres—were hugely impressive. The environmental movement felt in safe hands.

It was a rare high point on the international diplomatic calendar before the wave of populism that gripped the US, UK, Australia, Brazil, India and elsewhere in the months and years that followed.

I asked Nathan Fabian how he felt after the COP 21 agreement. "My sense of emotion was partial vindication, partial relief that we had been able to agree on what the problem was and what level of response would be needed."

"It wasn't a personal vindication. Rather, it was for the people that had been working for 30 years on climate change, that we'd finally got the right international recognition of the problem. Very late as it was, it nevertheless created the starting point we needed for real action."

"I was much less confident that we had the framework that was going to achieve the outcome. But it was essential that we first agreed on what the problem was to build the many years of additional work that was then required."

I asked Philippe Zaouati, founder and CEO of Mirova, the same question. "It was really a very emotional time. Mirova had existed for just 3 years."

"We had the 2009 Copenhagen summit in our minds. We went to Copenhagen just after the summit. It was in Scandinavia and we thought there would be a very strong outcome, but we were very disappointed."

"So we had lowered our expectations for COP 21, but the final agreement was strong and COP 21 was a real success."

"And to accompany the COP, we had this climate finance day just a couple of weeks before, where all the financial players, in particular in Paris, made a series of strong commitments. AXA said it wasn't possible to insure a world at 4 degrees Centigrade, a number of asset managers said they would stop investing in coal."

"When Laurent Fabius finalized the Paris Climate Agreement with the green hammer it was a very emotional moment."

For responsible investment, COP 21 gave rise to a number of climate change disclosure requirements, some of which are more successful than others. I'll start with France's Article 173.

Article 173

For French investors, one significant outcome of the Paris Climate Agreement was The Energy Transition Law. At the time, the regulation was groundbreaking.

With Brexit and Trump on one side and the Paris Climate Agreement and Article 173 on the other side, so began a period of French leadership on responsible investment—one that has prevailed until today.

Article 173 of the French Energy Transition for Green Growth Law was the first serious attempt by policymakers to require investors to manage climate change-related risks, and indeed, to decarbonise their investments (Republique Francaise 2015).

The Law was published in August 2015, ahead of COP 21, and applied to listed companies, banks and investors. The Law required companies to disclose in their annual reports the financial risks associated with climate change and the consequences of climate change on a company's activities, goods and services.

Banks were required to undertake stress tests on climate change-related risks. Investors would have to disclose how ESG criteria were used in investment decision-making and how their investment policies align with the aims of the energy transition.

The law covers both climate change and ESG issues. For climate change, it includes both transition and physical risks, and, importantly, an assessment of the contribution towards meeting climate change targets.

It also requires an asset manager to disclose how they consider ESG issues in investment decision-making, and, again importantly, the percentage share of funds that integrate ESG criteria.

With the benefit of hindsight, the disclosure was mostly narrative-based and not well enforced. In a follow-up study the PRI concluded, "Article 173 of France's Energy Transition for Green Growth Law has not yet achieved its original intent. Following policymaker consultation with investors, the law was introduced on a "comply or explain" basis."

"There is, however, no further guidance or agreement about the expectation of what would be a satisfactory explanation for non-compliance" (PRI 2018).

For its time, the law was groundbreaking. It was one of the first attempts by policymakers to enhance and formalise sustainability-related disclosures. Subsequent European sustainable finance regulation evidently has its roots in Article 173.

Philippe Zaouati agrees. "Article 173 of the French Energy Transition Law was the catalyst that led to the European policymaking process on sustainable finance because it was the first one."

"Sometimes when you set a new regulation, those in the finance industry say they are concerned by the regulation, they say it's a bad thing, and that it will lead to competitive disadvantage. But with Article 173, it was exactly the other way round. For French asset managers, it was a big competitive advantage."

"I remember when I was travelling in Europe and the US, everyone was speaking about Article 173. It was a way for French asset managers to start being ahead of the curve."

"I remember when I was a member of the HLEG in 2017 the French regulation was the starting point of the discussions."

"And Article 173 was clearly a consequence of COP 21. Because COP 21 was organised in France, Ségolène Royal wanted to showcase something at the COP that could be differentiating for France, and this was it."

Implementation was key to Article 173's success and advocates of the regulation deserve credit. Despite challenges with its implementation, it helped French investors leapfrog their Australian, British, Californian and Canadian counterparts as the maillot jaune of responsible investment.

One year after the publication of Article 173, I found myself at a conference in Paris, talking about implementation, which, prompting a few awkward laughs, the organisers had shortened to "clim-acts".

TCFD

Hot on the heels of Article 173 was TCFD. TCFD stands for the Task Force on Climate-related Financial Disclosures.

The Task Force was set up in 2015 by the Financial Stability Board (FSB) to standardise and harmonise climate disclosures (FSB 2015). Some companies and some banks were disclosing climate-related risks, but the disclosures were not comparable and not decision-useful. Much of the disclosure was marketing.

The Task Force was the coming together of two personalities, Mark Carney and Michael Bloomberg, supported by banks, investors and academics. The PRI seconded a member of staff to the Task Force.

It was deliberately private sector led. There was pushback against any mention of regulation for fear of derailing the Task Force before it had started.

Until recently, the Task Force still existed, superseded by the International Sustainability Standards Board (ISSB). Companies, banks, investors and service providers can sign up as a TCFD supporter. At the time of writing, there are over 3,000 supporters across 95 jurisdictions.

TCFD is a transparency initiative. The TCFD website says, "The FSB created the TCFD to develop recommendations on the types of information that companies should disclose to support investors, lenders, and insurance underwriters in appropriately assessing and pricing a specific set of risks— risks related to climate change" (FSB 2015). The TCFD is primarily aimed at companies.

Also on the website, in speech marks alongside a picture of "Michael R. Bloomberg" is the quote, "Increasing transparency makes markets more efficient and economies more stable and resilient."

To support its work, the TCFD hosted workshops, webinars and consultations. There were "more than 200 responses" to the first consultation.

I was privy to the Task Force's review of the consultation responses. The review of the written consultation responses was not thorough (at least, that was my interpretation). Given Bloomberg's network, and that of senior members of the Task Force, my insights at the time were that there was a fairly clear idea prior to the consultation on what the recommendations would look like.

The extent of consultations on responsible investment topics is considerable. Since my TCFD experience, I've only ever paid lip-service to such consultations. If you want to influence these frameworks you have to get to know the decision-makers.

The recommendations can be visualised as four concentric circles representing four themes: governance, strategy, risk management and metrics and targets.

The final report, published in June 2017, is 74 pages long, and introduces topics familiar to most investors such as scenario analysis (FSB 2017).

1. Governance: Disclose the organisation's governance around climate change-related risks and opportunities, this includes the extent of board oversight and role of management, as well as how decision-makers inform themselves on climate change topics.
2. Strategy: Disclose the actual and forward-looking risks related to climate change on business strategy and financial planning. Given the uncertainty of climate change, this incorporates scenario analysis. Scenario analysis measures the expected loss to the company across different temperature scenarios. In other words, the loss to the company at, say, 1.5 degrees degrees Centigrade, 2 degrees Centigrade or 3 degrees Centigrade, including both transition and physical risks.
3. Risk management: Disclose how the organisation identifies, assesses and manages climate-related risks, and how these processes are incorporated into management decision-making.
4. Metrics and targets: Disclose Scope 1, 2 and if appropriate (TCFD's words, not mine), Scope 3 GHG emissions, alongside the targets used to manage climate-related risks, and progress towards those targets.

TCFD is remarkably successful. Most large listed companies in Europe, UK, US, Canada, Australia and perhaps Japan (and elsewhere) will have some form of climate change policy, an understanding of the TCFD framework, and have started to measure and disclose emissions-related metrics. In many countries, including soon the US, TCFD reporting has become a regulatory requirement.

TCFD is a financial risk framework, not a real-world impact framework.

TCFD has put climate change risks and opportunities firmly on the agenda of corporate boards, pension scheme trustees and management teams at asset managers, in a way that no other framework has achieved.

The externalities of burning fossil fuels are largely unpriced risks in financial markets and TCFD is changing that. TCFD is arguably one of the biggest successes of the responsible investment industry.

Scenarios, Metrics and Targets for Pension Funds

The TCFD regulations published by the UK's Department for Work and Pensions (DWP) were in three iterations. Other jurisdictions have closely followed the UK's approach, but I'll focus on the UK as it was one of the first countries in the world to formalise TCFD reporting for pension funds.

First, the DWP required trustees' beliefs on climate change to be incorporated in the Statement of Investment Principles (SIP). The SIP is a scheme's primary governance publication.

Second, the DWP required trustees to publish a TCFD report, with the largest pension schemes publishing first, followed by smaller pension schemes.

The DWP's TCFD reporting requires at least two climate change scenarios, one of which must be "Paris-aligned" or "well-below 2 degrees Centigrade", as well as three emissions metrics, which must include an absolute emissions metric, an emissions intensity metric (also known as carbon footprint), and one other metric at the Trustees' discretion, and then a target.

The target should apply to one of the metrics. It does not need to be a net zero target, although many schemes have indeed set a net zero target, perhaps influenced by NGOs such as Make My Money Matter.

A UK industry group, the Pensions Climate Risk Industry Group (PCRIG) published guidance for pension schemes on TCFD reporting. In practice, the metrics disclosed by pension schemes are the result of a combination of regulation, industry group guidance (including IIGCC's Paris Aligned Investment Initiative), and the data made available by service providers, such as MSCI or Sustainalytics.

Third, the DWP required trustees to publish an alignment metric.

The most complicated aspects of TCFD reporting are scenarios and metrics.

Scenarios measure the financial loss to a company, portfolio, or portfolios at a given degree of temperature rise. The loss, disclosed as a percentage, incorporates both the physical risks of climate change and the transition or policy risks of climate change. Some models incorporate potential legal risks or the upside of environmental solutions.

The first set of models, tended to include scenarios at 2, 3 and even 4 degrees Centigrade of warming. The IPCC and others have since set out the extreme severity of the physical effects of 4 degrees Centigrade, and more recent models tend to include 1.5, 2 and 3 degree scenarios. Even 3 degrees Centigrade is so extreme as to be unlikely.

To measure the physical risks, models map companies' supply chains, operations and consumer-base by geography, and overlay with expected changes to weather patterns. An agricultural company, for example, with farmland in areas of water stress, is likely more financially affected than farmland elsewhere, and so on. The models incorporate issues such as drought, flooding, coastal erosion, extreme wind and wild fires by location and revenue.

The models tend not to incorporate second-order effects, such as famine or mass migration, nor incorporate environmental tipping points, nor model climate change beyond around 15 years or so. As such, the results tend to under-report the financial effects of climate change. As such, prevailing approaches to scenarios have been criticised for at best, not being decision-useful, and at worst, being misleading.

To measure the transition risks, some models assume a carbon price, other models a more granular mapping comparing existing public policies with the public policies necessary to limit warming to the selected scenario.

A lower degree of warming would require a nearer term and more comprehensive policy response (or a higher carbon price), which would affect company profitability. To meet 1.5 degrees Centigrade, for example, a company would need to retrofit its offices with improved energy efficiency, replace its car fleet with electric vehicles, perhaps replace components in its supply chain with more expensive, lower carbon equivalents—all at cost to the company.

To determine the overall loss, the model sums both physical risks and transition risks. For investors, it does so for each company in a portfolio. At portfolio level, there are some attempts to consider implications through supply chains, although this is mostly work in progress.

As one pension fund trustee put it to me, "The results provide spurious accuracy." A dollar loss, to the cent, based on a series of assumptions, which are almost certainly wrong, and in aggregate, materially so. But rather than an accurate prediction of the costs, the scenarios are intended to prompt actions. They're intended to be a starting point.

The scenarios may help companies in their climate adaptation, moving operations to areas less exposed to weather events, or may help companies explain to their investors why they're bringing forward capital expenditure to reduce emissions.

The two primary metrics of the DWP requirements are absolute emissions metrics and emissions intensity metrics. Again, this approach has been replicated in other jurisdictions.

There are various ways to attribute the emissions of companies to that of investors. One is to determine the emissions per unit of sales, the other per unit of value. The first goes by the acronym WACI (weighted average carbon intensity). WACI calculates the tons of GHG emissions per unit of sales, say per $1 million of sales.

While the C of WACI stands for carbon, we tend to interpret as carbon dioxide equivalent, in other words, incorporating all greenhouse gases.

WACI measures efficiency. The more efficient the company, the less the GHG emissions per unit of sales. It allows investors to allocate to more efficient companies. As a measurement, WACI tends to be more prevalent in the US, although many UK and European investors use WACI too.

Another way is per unit of value, and here the acronym is EVIC (enterprise value including cash). EVIC calculates the tons of GHG emissions per unit of value, both market capitalisation and issued debt (and, cash, although for most companies, cash is included in market capitalisation).

Market capitalisation is the company's traded equity value (the sum of the value of the company's shares). We add issued debt so that we have an enterprise value. This also allows investors to consider emissions across both equity and credit portfolios.

Let's say the company has £100 of equity and £50 of issued debt. Let's say I own £1 of equity and £2 of debt. As such, I own £3 of £150, or 2%. Let's say the company's annual emissions are 500 tons of GHGs. As such, I "own" (or am responsible for) 2% of 500, which is 10 tons.

EVIC tends to be more prevalent in the EU, and is the preferred metric of IIGCC's Paris Aligned Investment Initiative. EVIC attributes the emissions based on ownership, and as such, is often called "financed emissions".

When determining absolute emissions, we simply sum the emissions associated with each company in which we invest across our portfolios. While an important metric, the result is entirely dependent on the size of the portfolio. The bigger the investor's assets under management, the bigger the investor's absolute emissions.

As such, the DWP also requires pension funds to disclose emissions intensity, which it calls "carbon footprint". There are various definitions of carbon footprint. Carbon footprint is the amount of carbon dioxide equivalent, or greenhouse gas, emitted by an activity, individual, company or government, or in this case, the portfolio per unit of investment.

Emissions intensity simply divides the absolute emissions by the total value of the portfolio (and typically, multiplies by $1 million). As such, it gives us the GHG emissions of the portfolio per $1 million invested.

This metric is also subject to market fluctuations. Companies with high emissions may increase in value, and in doing so, become a larger proportion of the portfolio, and so emissions will increase even if the investor has made no changes to their investment.

However, it is a much more comparable metric than absolute emissions. It allows for comparison across fund and across time series, and is the metric most commonly used for target setting.

The scenarios and metrics relate to corporate equity and credit. Theoretically, it allows investors to add together their emissions across multiple investment strategies, in both public and private markets, in equity and credit.

However, most TCFD reports disclose by strategy. The emissions intensity of an emerging markets portfolio is almost certainly higher than that of a developed markets portfolio. The emissions intensity of a private equity portfolio almost certainly uses a number of assumptions, as disclosure is less prevalent than listed markets.

The metrics do not work for sovereign investments, nor commodity investments. They do work for derivatives, such as equity futures. However, technically, in a future, the transaction is speculation on a price, not financed emissions (the investor is not financing the company, rather it is entering into a swap with a counterparty based on a predetermined price). However, TCFD measures the financial risks of climate change, and the value of the future, in theory at least, should be affected by climate change-related risks in the same way as the physical equivalent.

For sovereign investments, some investors apply the same approach as for corporate financed emissions, whereby, the investor calculates the percentage ownership of issued government debt.

Let's say the government issues £100 of debt, and I own £5, then I'm responsible for 5% of the government's emissions. This however favours countries with large debts. Let's say Germany has issued £100 of debt and Japan £1000 of debt, and I own £5 of each. I'm responsible for 5% of Germany's emissions, but only 0.5% of Japan's.

This is nonsensical, and so most investors tend to change denominator, either to GDP or population. GDP gives us a measure of a country's efficiency, in other words, the emissions per unit of GDP. This tends to favour developed countries with larger GDPs. Another measure, and the one I favour, is per size of population. This is still unfair. It doesn't incorporate a country's historic emissions. But it's the least worst, in that, a ton of GHG emissions has the same contribution to climate change, regardless of where it is emitted.

At the time of writing, there is no established sensible methodology for the GHG emissions of a commodity investment.

Climate Tilts

In late 2016, the defined contribution pension fund for HSBC staff committed to invest in a climate change tilt. When HSBC pension fund made its net zero commitment (in October 2021), the press release states that the scheme's trustee "was the first to include a positive climate risk tilt in its DC default fund." It was the first to do so for the default pension scheme, rather than an "opt in" green option.

Climate tilts work as follows.

First, the investor takes a benchmark, in HSBC pension fund's case, it was the FTSE All-World equities index. Next, the index is revised, solving to a decarbonisation or green revenues objective, by over-weighting or under-weighting index constituents.

Scope 1 and 2 GHG emissions can be reduced by as much as 50% by removing several of the most polluting companies, over-weighting lower carbon companies, and indeed, lower carbon sectors or geographies.

Tracking error, a metric that determines how closely the revised "climate index" tracks its parent index, can normally be limited to just a per cent or two.

HSBC pension fund's CIO described the approach as the "new normal". It was ahead of its time, and a few years later, both workplace and auto-enrolment pension funds were following suit.

The EU has since established two benchmarks, a climate transition benchmark, which requires a 30% emissions reduction (from the parent index), and a Paris-aligned benchmark, which requires a 50% emissions reduction.

At the time, it was a significant step forward for HSBC pension fund, FTSE and LGIM (the fund manager that invested the lower carbon index).

In 2020, as economies came to a standstill in response to Covid-related lockdowns, the investment approach outperformed. Pharmaceutical and technology stocks outperformed, as we consumed less energy and watched more Netflix.

In 2022, that was reversed, as industry, airlines and other high carbon sectors returned to full-strength, and Russia's invasion of Ukraine disrupted energy markets, and pushed up energy prices.

As well as higher carbon companies, some tilts also excluded defence, another sector that outperformed in 2022, as Ukraine's allies both armed Ukraine and themselves.

Potentially, the tilts help manage the financial risks associated with unpriced GHG emissions. Companies with higher GHG emissions will be disproportionately affected by the rising cost of emissions (assuming governments take action to price the externality).

The more investors that apply tilts, the more high polluting companies will be excluded from investment portfolios. This (theoretically) will push up their cost of capital and challenge their ability to compete with their lower carbon competitors.

Tilts, however, should be handled with caution.

Across an index of several hundred companies, many large in size, and companies likely far more affected by government policy, consumer action or geopolitics, for most companies, a tilt towards or against is unlikely to substantially affect their cost of capital, and force an under-weighted company to chart a new, more environmental course.

Indeed, the tilts are often based on Scope 1 and 2 emissions, and in a global index, decarbonisation can be achieved by over-weighting developed-market economies, and under-weighting emerging-market economies, which from a sustainability point of view, is counterproductive, denying capital to the economies that most require it to transition.

Some argue that it is not cost of capital, but market signal that matters. This resonates with me, but there is little evidence to suggest this is the case, and if I had to guess, I'd say it's wishful thinking.

In late 2016, it was a step forward. And tilts have their place—even today. But we need to be careful not to overstate their real-world sustainability impact.

"People have been tilting funds with climate factors for at least 20 years now," Nick Robins said to me.

"The key is to understand what part of the climate challenge is being addressed."

"At Henderson [since its merger with Janus Capital, known as Janus Henderson], we undertook the first carbon footprint of our SRI funds back in 2005. We found that our Industries of the Future green fund, which was investing in climate solutions, came out with a higher carbon footprint than our income fund."

"The reason? We were only measuring Scope 1 and 2 emissions and so were counting the emissions from industrial companies (using a fossil intensive energy system) and not balancing this with the emission reductions that these companies generated in Scope 3 via the products they sold."

"And it's important to recognise that different tools can give you different answers: carbon footprints give you a picture of emissions performance. But look carefully at what scopes they cover. Carbon footprints don't tell us about risk. More fundamentally, a lot of emission based climate tilting is based on past performance; it tells you where the portfolio was yesterday."

"Nearly all investment is about future performance and this is where climate-tilting is heading, understanding how the assets in a portfolio are aligned with the 1.5 degrees Centigrade global temperature target. This means a much closer examination of transition plans which provide this forward look."

"Crucially, the key thing for investors to understand is how companies are positioned for making a transformational shift in climate performance by 2030 by which time global emissions need to have almost halved from 1990 levels. This focus on the next 1, 3, 5, 7 years is much more aligned with investment time-frames than far-off 2050 targets. There needs to be a real focus this decade. This is the swing decade."

Alignment Metrics

Alongside emissions metrics, we have started to see the emergence of alignment metrics.

Alignment metrics measure the extent to which the portfolio is aligned with the Paris Climate Agreement. Inevitably, this is more complex than it sounds.

In October 2021, the DWP set out its plans to require UK pension schemes to disclose an alignment metric. The DWP's finalised plans were published in June 2022, all UK pension schemes, with more than £1 billion in assets must disclose an alignment metric (DWP 2022).

Two weeks after the DWP's initial announcement in 2021, the TCFD portfolio alignment team (PAT) published its assessment of approaches to portfolio alignment (TCFD Hub 2021). In August 2022, the GFANZ published a consultation paper setting out four approaches (GFANZ 2022).

The four approaches range in complexity, starting with a binary target metric. A binary target metric measures the percentage of companies, typically by value, within a portfolio that has a net zero target, that is independently verified, often by the Science-Based Targets initiative (SBTi).

An extension of the binary target metric is the "maturity scale" metric. Here, rather than a "yes no" on a net zero target, the metric is the percentage of companies across a series of net zero categories. For example, companies that have set a target that is 2 degrees Centigrade aligned (but not 1.5 degrees Centigrade aligned), companies that have a set a target that is 1.5 degrees Centigrade aligned (but not independently verified), companies that have set a target that is 1.5 degrees Centigrade aligned and independently verified and so forth. Personally, I find this metric decision-useful, although the categorisation needs to be trustworthy.

A third metric, and one that has received a lot of attention from the service provider and asset manager community, is an Implied Temperature Rise metric (ITR). An Implied Temperature Rise metric is expressed as a degree of warming, say 2.5 degrees Centigrade. It tells us that the portfolio is consistent with (or contributes to) 2.5 degrees Centigrade of warming.

The final metric is the benchmark divergence metric, which tells us how much the portfolio diverges from an established benchmark, say the EU's Paris-aligned benchmark (PAB), which requires 50% emissions reduction (compared to the mainstream equivalent benchmark), and 7% emissions reduction per year.

The ITR is compelling in its simplicity. A trustee of a pension scheme may take comfort in knowing that their portfolio is well below 2 degrees Centigrade aligned. In practice, there is a range of assumptions involved in determining the portfolio's implied temperature.

Two of the most useful approaches that I've reviewed are that by MSCI and by asset manager, Lombard Odier. Here is my understanding of both:

MSCI uses a range of inputs, including existing carbon budget and countries' nationally determined contributions by sector to determine a unique pathway for a company's Scope 1, 2 and 3 emissions (MSCI 2022).

MSCI then determines the company's divergence from that pathway, either undershoot (positive) or overshoot (negative). The divergence is translated from an emissions metric into a temperature metric. This is calculated across a portfolio to give a portfolio's implied temperature.

This approach has a series of advantages. Its starting point is the remaining carbon budget and nationally determined contributions, which is transparent. Because it is unique to each company, it (in theory) ensures that investors cannot manipulate portfolios by sector to lower the implied temperature (for example, overweight low carbon sectors and underweight high carbon sectors). It does however require a range of assumptions, which, while thoughtful, are largely arbitrary.

Lombard Odier categorises companies in material sectors into "ice cubes" and "burning logs" (Lombard Odier 2021). Ice cubes are companies that reduce global warming. These companies operate well below the GHG emissions expected for their sector and geography. Burning logs are the opposite. These companies operate well above the GHG emissions expected for their sector.

To lower the portfolio's implied temperature, Lombard Odier tilts away from burning logs in favour of ice cubes.

This approach allows Lombard Odier to invest across a range of industries, over-weighting the companies that are best performing (or most aligned) from a climate change perspective, but not excluding whole sectors.

The alignment metrics are helpful and MSCI and Lombard Odier set out good practice, but we should consider their use in portfolio construction as a form of advanced climate tilts.

The Rise and Rise of Net Zero

Since Article 173, TCFD, and a range of other regulatory disclosure initiatives, the big climate theme moving into 2021 was "net zero by 2050" or "net zero" for short. That investors—and the companies we finance—are not adding to the amount of GHG emissions.

To limit global warming to 1.5 degrees Centigrade, as set out in the Paris Climate Agreement, we need to get to net zero GHG emissions by 2050. The word "net" is important. Total emissions need to be zero, offset by new carbon capture and storage technologies, include nature-based solutions.

Net zero has given way to a suite of terms, such as net zero aligned and science-based targets.

A temperature target (here, 1.5 degrees Centigrade) translates into a remaining carbon budget (an amount of GHG emissions that we can emit that allows us to limit warming to 1.5 degrees Centigrade, normally understood to be by 2100). This carbon budget is then applied across sector and geography in order to provide us with a pathway that can be reasonably expected for any given company's decarbonisation strategy. We can then test a company's actual decarbonisation strategy against this pathway to see whether it is net zero aligned.

Some sectors that are easier to transition—where the technology exists— such as energy production will have a steeper decarbonisation trajectory than sectors that are hard to transition (such as cement or steel).

Of course, there are considerable dependency issues here. If one sector fails to transition at the pace it needs to, then other sectors must overcompensate and transition more quickly.

For now, we tend to apply net zero targets to just Scope 1 and 2 emissions, which includes the emissions associated with the direct burning of fossil fuels or the emissions associated with the burning of fossil fuels to provide electricity or heat.

For most firms, particularly in higher income countries, the majority of emissions are Scope 3 emissions, which are the emissions associated with "upstream" company supply chains and the emissions associated with "downstream" use of the company's products by its customers.

When we assess a net zero target of a company or an investor, we need to know the shape of the decarbonisation curve. What's the starting point? What's the interim target? And does the target apply to Scope 3?

Another important issue in the assessment of net zero targets is the use of offsets. Offsets can be controversial. Offsets are unregulated, mostly in lower income countries, with huge variation in estimated future prices. Companies, such as airlines, with no proven net zero technologies, can claim to be net zero aligned by incorporating largely under-valued offsets in their decarbonisation plans.

Research by Trove Research found that to meet their net zero targets, companies "have promised to plant more trees by 2050 than there is space" to do so (The Times 2023).

But while there are many pitfalls to be avoided with net zero, it is at least a framework, and in theory at least, the framework is rooted in science.

Sofia Bartholdy set out the Church Commissioners for England's approach to net zero.

"Our net zero target is a target to achieve a net zero emissions portfolio by 2050, and importantly, align with a net zero world."

"The reason we have a net zero target is because we believe limiting the global temperature increase to 1.5°C is the best outcome for people, planet and the economy. We, therefore, want to ensure that our approach to achieving our net zero target focuses on how we can be part of the solution and use our levers, as an investor, asset owner and the Church Commissioners specifically, to influence outcomes and move towards a net zero world, not just a net zero portfolio."

"We see our levers for change in broadly three categories; aligning our processes, engaging and investing in solutions."

"Aligning our processes – these are actions that in themselves will not have an impact, but if enough investors do the same, it may cause a ripple effect. Our public net zero target, for example, is part of providing a social licence for policymakers, companies, and other investors to align with a net zero pathway. Our climate-related exclusions may also not change the behaviour of companies by itself but is a signal to the market that we have lost faith in companies operating in certain activities' willingness and ability to transition."

"Engaging – we believe in engagement as a key tool for change. We use our position to engage with various types of stakeholders from our farming tenants, to large listed companies our asset managers and governments on their approach to the climate crisis."

"Investing in solutions – we have an opportunity to invest in the energy transition and in climate resilience. This can take different shapes across the portfolio and is a key lever and crucial for the success of the energy transition."

"We believe that there is an overwhelming likelihood that the energy transition will happen. Whether it happens fast enough to meet the goals of the Paris Agreement and whether it is just or nature positive, are big question marks. We continue to work on a holistic approach, in which we consider the interconnected nature of the global economy, and where we use our available levers to influence change."

Martin Spolc, Head of Unit for Sustainable Finance at the European Commission said, "The end game is that, by 2050, climate neutrality must be achieved. We don't have an alternative."

"To get there and to ensure that companies and the financial sector are aligned with our longer-term targets, I think that it's always more useful to talk about what needs to happen over the next 5 to 10 years."

"During that period, first, it would be good to have a clearer and better understanding of where the sector is. We don't currently have such a clear understanding. We have developed rules on disclosures, but my first wish is that we have a better understanding of how effective these transparency rules and the tools that we have developed have been."

"Second, it would be good to have a clearer and better understanding of the risks. To understand better the severity of the risks we are facing. So far, the evidence suggests that the risks are getting worse and can interact in a rather unpredictable way, reinforcing each other."

"Third, it would be good to have a system and governance in place that would allow policymakers to adjust their policies and use a more flexible dynamic policy toolkit based on the findings of the first two pillars. First, whether and to what extent the financial sector and companies are aligned with the net zero trajectory and second, the degree and nature of risks that we all will be facing."

"Finally, I wish that over the next 5 to 10 years the sustainable finance agenda becomes truly mainstream, despite the fact that not everybody will be fully aligned yet. The challenge with transitioning is huge. That requires massive efforts from everyone. It will be important that stakeholders will have developed credible pathways and transition plans on how to get aligned over time."

"In any case, it's important that we've made enough progress by then, because if we are too slow in the beginning of the transition, then we will have to be much faster towards the end, which will be likely much more costly."

COP 26

The backdrop to COP 26 was net zero.

In 2019, the UK Government committed to net zero. In a press release, the government said "The UK today became the first major economy in the world to pass laws to end its contribution to global warming by 2050" (UK Government 2019).

Other governments followed suit.

Also in 2019, the EU published its Green Deal titled, "Striving to be the first climate-neutral continent" (European Commission 2019).

As did companies.

In 2020, TotalEnergies set its target. "Net Zero across Total's worldwide operations by 2050 or sooner (Scope 1 + 2)" and "Net Zero across all its production and energy products used by its customers in Europe by 2050 or sooner (Scope 1+2+3)" (TotalEnergies 2020).

Also in 2020, the Oneworld Alliance, which includes American Airlines, Japan Airlines, Cathay Pacific and British Airways set its net zero target. American Airlines said it will reach the target "through various initiatives such as efficiency measures, investments in sustainable aviation fuels and more fuel-efficient aircraft and carbon offsets, among other measures" (American Airlines 2020).

By November 2021, the date of COP 26, governments, companies and investors were committing to net zero.

COP 26 was hosted in Glasgow, UK. A curious mix of NGO staffers, private sector sustainability teams and overseas policymakers sported an array of lanyards, all walking with phone in hand, between Glasgow Queen Street train station and the secure zone's metal barriers about a mile or so away. The roads in between guarded both by the police and a handful of committed activists, literally and figuratively banging the drum.

COP 26 was significant because it marked the (delayed a year due to Covid) five-year anniversary of COP 21. COP 26 required every government to submit their contribution to decarbonisation, their NDC, or "nationally determined contribution". As such, it was perhaps the most important barometer we have had on progress towards tackling climate change.

Glasgow is cut in half by the M8 motorway, but it's a city with charm. Its gothic architecture, river bank paths, museums and parks are well worth a visit. The surrounding countryside is majestic. The city's history is remarkable. Ardgowan Castle, a short drive to the West, is as beautiful as it sounds. And the city has grit. If you haven't yet read Shuggie Bain, you should.

But, Glasgow just wasn't a big enough city to host a conference badged "make-or-break" for the future of humanity. For one, there just weren't enough hotels. One asset manager ferried their staff back and forth from Edinburgh in a taxi. I too stayed in Edinburgh. I took the train.

NGOs hosted well-meaning events attended by a handful of their supporters, often staff at other NGOs. The NGOs' comms professionals posted photos to their social media accounts, often commuting into Glasgow from a sofa bed at an overpriced, out-of-town Airbnb.

Access to political leaders was tightly controlled. Thanks to Lombard Odier, I briefly met then Prince, now King Charles. King Charles was far more open to conversation than political leaders.

Never failing to disappoint, the star of the show was Greta Thurnberg. COP 26 was home to Thurnberg's famous "blah blah blah" speech at the pre-COP Youth4Climate summit. Thousands, and it may have been tens of thousands, but it certainly wasn't hundreds of thousands, marched the streets for a "global day of action".

Alok Sharma, the COP president, was, without doubt, a shining star in an otherwise disappointing UK Government line up. One government insider who, for obvious reasons, asked not to be named told me, "the government put forward its B team." She went on, "the A team has spent their year flying between London, Brussels and Dublin working on Brexit."

UK Prime Minister Boris Johnson barely turned up. He flew of course. He sat next to broadcaster and environmentalist David Attenborough without wearing a mask. He snoozed. He flew back to London for a fundraiser. He made it clear that he just wasn't interested.

But there were some reasons to be optimistic.

The commitments to cut methane emissions, end deforestation by 2030, and reduce reliance on fossil fuels were important steps forward.

And it was not just the usual suspects; significant commitments from India, Brazil and Russia, with India committing to 500 gigawatts of renewable energy by 2030 and Brazil and Russia joining the commitments to end deforestation (COP 26 2021).

But, overall, the announcements fell far behind what's needed.

At the conclusion of the conference, a finalised 10-page negotiated outcome document titled the Glasgow Climate Pact was released. This was the first ever climate deal to explicitly plan to reduce coal, the most polluting fossil fuel.

However, the agreement only promised to "phase down" rather than "phase out" coal, amid statements of disappointment by some.

The World Climate Summit, which took place alongside COP 26, was unfortunately part of the problem. Badged, "the investment COP", it was a pay-to-play conference. It seemed to me that it was not commitment to sustainability, or actions undertaken, that bagged you a speaking slot, but rather how much you were prepared to pay.

The big asset managers put forward an ESG professional for panel discussions or for the very deep pocketed, a key note. Despite the cost of the conference, the presentations were, on the whole, disappointing. One US key note speaker started by saying "I believe in climate change."

There was no feet-to-the-fire here, perhaps, because the speakers had paid to speak. Little to learn. No chance for genuine collaboration or problem solving on challenging topics.

It was a missed opportunity.

For the first time in two-years (due to Covid), Glasgow was host to NGOs, companies, investors and policymakers, but each in their own bubble, the continuation of an approach that wasn't working, and won't work.

But despite the disappointment of the investment COP, COP 26 did mark the launch of GFANZ, pushing net zero to the top of investors' agendas.

I decided not to attend COP 27, hosted in Sharm El-Sheikh, one year later.

COPs provide us with an inflection point—a calendar entry for governments, companies and investors to reflect on their progress and further their commitments. That's ok, awareness raising has its place. But flying to Sharm El-Sheikh—to be a bystander—is less compelling than a train ride to Paris or Glasgow.

The next significant COP is 31, where governments will put forward their next round of NDCs.

Green Bonds

Green bonds are bonds in which the use of proceeds (what the bond is financing) achieve an environmental objective.

As ICMA, home of the Green Bond Principles, says, "Green bonds enable capital-raising and investment for new and existing projects with environmental benefits."

The Principles set out four components, "1. Use of proceeds, 2. Process for project evaluation and selection, 3. Management of proceeds, 4. Reporting" (ICMA 2021).

To be investable, green bonds will undergo a third-party review with a number of service providers charging the parent company (or government) to do so.

When investing in green bonds, it's all about the green premium, or greenium. All things being equal, green bonds are more expensive than non-green (or traditional) bonds.

This is mostly because green bonds require additional layers of reporting, independent verification (to assess that the bond is indeed green) and potentially additional layers of regulatory disclosure.

The credit risk is, however, unchanged. If the company (or government) defaults, there is no special treatment for green bonds.

Green bonds are, however, an attractive asset class and in recent years, the green bond market has ballooned. And so, even though green bonds can be more expensive, they can also out-perform on the secondary markets.

Bonds, unlike equities, can be used to finance a specific part of a company's operations. A green bond could, for example, finance the retrofitting of buildings, low carbon transport or renewable energy. Investors will need to be comfortable that the use of proceeds meets their environmental objectives and that the greenium meets their risk and return objectives.

Most green bonds are oversubscribed (there is more demand than supply). When a company issues a green bond, it would typically work with an investment bank (or even, multiple investment banks). The banks would, in turn, work with investors.

Investors would submit an order. Banks would allocate. In the case of a green bond, banks may have bias towards investors that have a track record on sustainability.

The bonds may be traded in the secondary market.

Companies will, on the whole, want to see their bond increase in price (more buyers than sellers) as it's an indication that investors perceive the company to be credit-worthy.

But the company is unlikely to mind too much who owns the bond. The company will pay a coupon (and return the original investment) to whoever owns the bond at the time. Unlike equity investments, bond investments do not come with voting rights at company AGMs.

The data for the GHG emissions of green bonds is not readily available (often this is Scope 4 emissions, the GHG emissions avoided through the financing activities of the green bond). But that is evolving. As are reporting frameworks to check that the green bond has been invested according to the prospectus.

In February 2023, European policymakers announced provisional agreement on the European Green Bond standard (EuGBs), officially adopted in October 2023 (European Council 2023).

"Under the provisional agreement" the announcement said, "all proceeds of EuGBs will need to be invested in economic activities that are aligned with the EU Taxonomy, provided the sectors concerned are already covered by it. For those sectors not yet covered by the EU Taxonomy and for certain very specific activities there will be a flexibility pocket of 15%."

The response was mixed, with some concerned that the EU Taxonomy was too limiting and could deter companies from issuing green bonds. Others set out concerns that longer-dated green bonds may currently comply with the EU Taxonomy, but not for the bond's duration, as the Taxonomy's performance thresholds are tightened.

Given bond investments are favoured by de-risking DB pension funds, green (and social and sustainable) bonds are an important tool for pension funds in meeting their net zero commitments.

Regulatory oversight of allocations is an important step forward.

Inevitable Policy Response

One of the more complex features of TCFD is scenario analysis. One of the more useful scenarios has been put forward by a programme of work titled, "The Inevitable Policy Response".

In 2020, PRI commissioned the Inevitable Policy Response (IPR) (the acronyms are unfortunate).

Given that financial returns will be driven by how well investors forecast the transition, IPR sets out to forecast climate change-related policymaking, such as changing energy systems or the roll-out of electric vehicles and associated technological shifts.

The IPR thesis is that governments will not continue to stand by and watch the physical damages of climate change accelerate. As weather events continue in their frequency and severity, voters will demand politicians act, and so governments will, inevitably, introduce policy change.

While legislation to address climate change may take years, markets may reprice polluting activities in days or even hours based on certain signals. Some assets will be stranded (there will be no buyers). Some parts of the market may price in an acceleration of policy change, while other parts will be reactive. Non-linear policy change will therefore cause market disruption.

Investors should prepare accordingly. This is circular. If investors prepare it changes the pricing dynamics. It makes policy change more likely. And so on.

The lobbying dynamic is key here. While oil and gas is the largest contributor it is difficult to see short-term change. Green lobbying is however increasing rapidly and it is easier when renewable energy economics are becoming more favourable.

The Inevitable Policy Response does not have a temperature target as it works forward from the drivers of policy and technology, but its forecast outcome is around 1.8 degrees Centigrade. Unlike the net zero commitments, it's not saying, "we must limit warming to 1.5 degrees Centigrade", because IPR forecasts that 1.5 degrees Centigrade is unlikely to happen.

Rather, it forecasts policy change across a range of polluting industries. When, for example, will California ban combustion engine vehicles? When and how will China's energy systems transition? What about land use in Brazil, the Congo or Indonesia?

The majority of IPR forecasts have already been proved right and indeed some areas have exceeded the forecasts such as electric vehicles.

The PRI website says, "IPR forecasts a continued acceleration in climate policy to 2025, driven, in part by the 2023 Paris Stocktake and the 2025 ratchet," requirements of governments' signatory to the Paris Climate Agreement.

"IPR assesses that those policy responses will be increasingly forceful, abrupt and disorderly and produces in-depth scenarios to assist investors in navigating the financial, market and real economy uncertainties inherent in climate transition" (PRI 2021).

The IPR team takes their analysis directly to the world's biggest investors.

IPR allows us to determine a scenario of what we think is likely to happen, how that will affect our investment portfolios and how we can prepare for it.

If 1.5 degrees Centigrade is our target and 3 degrees Centigrade is our worst-case "hothouse" scenario, then IPR is our forecast. While we may hope for 1.5 degrees Centigrade, IPR is what we think will happen.

There's a tension within PRI between asking investors to commit to 1.5 degrees Centigrade while forecasting 1.8 degrees Centigrade—remembering that climate change is not linear, and there will be a considerable increase in the frequency and severity of weather events at 1.8 degrees Centigrade, and certainly at 2 degrees Centigrade.

This tension is becoming more obvious now that the World Meteorological Organization (WMO) has suggested 1.5 degrees Centigrade will be reached by the late 2020s (WMO 2023). It would seem evident that investors will struggle to maintain their commitment to something that is proved to be impossible. Therefore, the challenge for PRI and indeed all the net zero alliances is how to maintain investor ambition while acknowledging that an overshoot past 1.5 degrees Centigrade is inevitable.

The additional challenge for investors is the redefining of portfolio emissions to include Scope 3 upstream. IPR actually forecasts that the OECD block will reach net zero targets as a whole by 2050 but the residual emissions will be in non-OECD countries where Scope 3 supply chain emissions will remain high. As we approach the point where 1.5 ambition seems out of reach, investors and their alliances need a new set of options if the entire collaborative network is to maintain its solidarity and cohesion.

The analysis underpinning IPR is well worth reviewing. The individuals that run IPR are industry insiders and their work is high quality. It is publicly available on PRI's website.

"For a lot of investors, IPR helped to crystalise the idea that there would be a volatile policy transition on climate risk at some point" Nathan Fabian told me. "It was previously very hard to understand that in a tangible way. IPR helped a lot of people think 'if policymakers are going to act and it's going to be late, then it's going to be volatile'. That's the key conclusion and the analysis has reached 1000s of investors."

"For the more granular application of IPR - reviewing policy changes by countries assessing the policy gaps - it's a much smaller group of signatories that have done this, perhaps in the 100s of investors. And it's only in the 10s of investors that have done their own modelling on the transition pathways with their own pricing assumptions."

"IPR is an analytical tool that overlayed a forecast on the idea of scenarios. Markets price on the basis of what they forecast will actually happen, and IPR provided a forecast that analysts could compare and contrast with."

Carbon Markets

Set up in 2005, the EU Emissions Trading System is the world's first established carbon emissions market and was the largest until the launch of China's ETS. It covers 30 countries and over 17,000 installations.

It is a market system based on a "cap and trade" scheme or put more simplistically "pay to pollute". It's a way to, as smoothly as can be, raise the cost of polluting activities to the benefit of non-polluting activities.

It works as follows.

There is a cap on overall emissions by installations covering around 40–45% of the EU's GHG emissions, which includes around 11,000 power and manufacturing plants.

The cap is reduced over time so that total emissions fall.

Allowances are distributed through auctions (mainly still in the power sector) and free allocations (to reflect the 'progressive transition' to auctioning).

If a company reduces its emissions it can keep the spare allowances for future needs or sell to another installation that is short of allowances.

Nearly everyone I've spoken to working in responsible investment is in support of a carbon tax. It is the fairest way to decarbonise. It internalises the externality. It uses market mechanisms to make sure more polluting activities cost more.

More polluting activities cost more and so, all things being equal, will see reduced demand, thus lowering the pollution.

Carbon markets are also investable, which ensures the efficient distribution of permits. Trading helps ensure emissions are cut where it is least costly to do so. And a higher carbon price spurs investment and innovation into low carbon technologies.

"I would love for there to be a real carbon tax that's transparent, assured and regulated" Jon Lukomnik told me. Most of my other interviewees agreed.

In practice, however, there are challenges.

In 2018, I was living and working in Paris during the Gilet Jaunes protests. I lived centrally on Rue Godot de Mauroy, a short walk from Opéra. Weekend after weekend my road was blocked by protesters, installing and setting alight roadblocks, protesting against rising fuel prices.

When it comes to a carbon tax, there are profound political implications. Take the energy costs for social housing tenants living in post-war, low quality, low efficiency housing. Who assumes the cost? Or, in the case of the Gilet Jaunes protests, the cost of driving, where, for rural communities, there is no public transport alternative.

A global, meaningful carbon price remains a worthwhile pursuit, accompanied by carbon border adjustment mechanisms (CBAM), such that non-participating jurisdictions undertaking polluting activities cannot undercut pricing.

One consequence of a carbon tax is likely to be public companies selling the most polluting parts of their business to private investors or even to governments. It is more likely the public sector than the private sector that will be able to re-train, and if necessary, re-house workers in industries that are no longer compatible with environmental goals, decommissioning outdated infrastructure.

A carbon tax and a just transition, therefore, go hand-in-hand.

Green, Not Greenwashing

No book on responsible investment is complete without addressing greenwashing.

It was no deforestation day. My then company's (former) comms agency wrote in with a suggested tweet. To paraphrase, the suggested tweet was something about how integrating ESG issues will address deforestation and perhaps even "end" deforestation.

I'm sure this is true of many a topic, but on responsible investment, there's often a tension between compliance and marketer. A fund that saves the world sounds much better than a fund that integrates financially material ESG issues, even if it's the latter that most investors do. We didn't tweet about deforestation.

The regulators agree.

In May 2022, German police raided (the use of the word raid is, I think, noteworthy, translated by media outlets from the German word "durchsuchung") the headquarters of Deutsche Bank and its asset manager DWS to investigate what the regulator called investment fraud. The raid started following claims by whistleblower Desiree Fixler.

Desiree Fixler was DWS's first chief sustainability officer. In a FT interview, Fixler said she'd understood she was tasked to "align the entire platform", saying there was a "disregard for accuracy" in their ESG assets under management (FT 2022).

"Internally we're making certain statements," which Desiree said were contradicted in external statements, which "was misinterpretation," which "if material, is securities fraud."

Fixler's contract was terminated in 2021. "Any regrets?" asked FT journalist Patrick Temple-West. "No" Fixler replied. The market has successfully mobilised trillions of dollars into "ESG investments", but "we've not moved the needle here."

"Do you feel vindicated at all?" "A little bit" Fixler said, citing a FT article that found that DWS marked down their ESG assets by 75%.

"A sustainability officer is not a marketing officer. A sustainability officer should be considered as a compliance officer."

The DWS case was and remains a wake-up call to the responsible investment industry. I expect DWS's approach was typical of other investors too.

Also in May 2022, the SEC fined BNY Mellon 1.5 million USDs for "misstatements and omissions" about ESG considerations in mutual funds (SEC 2022a).

"The SEC's order finds that, from July 2018 to September 2021, BNY Mellon Investment Adviser represented or implied in various statements that all investments in the funds had undergone an ESG quality review, even though that was not always the case. The order finds that numerous investments held by certain funds did not have an ESG quality review score as of the time of investment."

And in November 2022, the SEC fined Goldman Sachs Asset Management 4 million USDs for failing to follow its own ESG policies and procedures between April 2017 and June 2018 (SEC 2022b).

There are other examples.

I have some sympathy for investors here. In its first 15 years, responsible investment was an unregulated activity. As one colleague described it, "it's like the wild west out there." Investors could do and say basically what they wanted. There was a proliferation of terms, methodologies, metrics and scores. Impact sounds better than ESG. And ESG sounds better than no ESG. And so an arms race on sustainability continued apace.

More recently, regulators have cracked down, in the SEC's case, retrospectively. Claims must be real, actioned and evidenced. If the policy says the investor integrates ESG issues then the investor should integrate ESG issues.

When the UK FCA introduced its consultation into sustainability disclosure requirements, its rationale was to address greenwashing.

"Consumers must be able to trust sustainable investment products. Consumers reasonably expect these products to contribute to positive environmental or social outcomes. There are growing concerns that firms may be making exaggerated, misleading or unsubstantiated sustainability-related claims about their products; claims that don't stand up to closer scrutiny (so-called 'greenwashing')" (FCA 2022).

The next wave of greenwashing has got to be net zero-washing, for both companies and investors alike. Many companies have made net zero commitments where the technology simply doesn't exist or through carbon offsets that are not properly priced.

On ESG claims, to some extent, I think the regulators have opted for a "shot across the bows". In monetary terms, 4 million USDs is unlikely to be a concern to an asset manager the size of GSAM. But reputation matters, and investors will want to avoid regulatory investigation, and so behaviours have changed, and new regulatory intervention, such as SFDR, SDR and proposed SEC disclosure obligations will help to provide some clarity to the question, "what exactly is ESG?"

The regulators have intervened in order to coalesce around common standards. There's the risk that greenwashing undermines responsible investment. If asset managers can label their funds responsible, green or sustainable but continue to invest as they always have done, responsible investment will be short-lived.

And for the future of responsible investment, it's important that there is confidence in process and practice.

But, when I'm asked about greenwashing, I always conclude by saying that the greatest risk is that we fail to achieve our sustainability goals. Tackling greenwashing is a means, not an end.

For responsible investment professionals, it can also be frustrating that greenwashing is discussed as if overstatement or misselling is a feature of responsible investment. It isn't. It's a feature of investment in general. And so it affects responsible investment, but it is not unique to responsible investment.

References

American Airlines (2020), ESG Report. [online]. Available from: https://www.aa.com/content/images/customer-service/about-us/corporate-governance/aag-esg-report-2019-2020.pdf (Accessed, January 2023).

COP 26 (2021), COP 26 Outcomes. [online]. Available from: https://ukcop26.org/the-conference/cop26-outcomes/ (Accessed, January 2023).

DWP (2022), Proposed Amendments to the Statutory Guidance. [online]. Available from: https://www.gov.uk/government/consultations/climate-and-investment-reporting-setting-expectations-and-empowering-savers/proposed-amendments-to-the-statutory-guidance-governance-and-reporting-of-climate-change-risk-guidance-for-trustees-of-occupational-schemes (Accessed, January 2023).

European Commission (2019), A European Green Deal. [online]. Available from: https://commission.europa.eu/strategy-and-policy/priorities-2019-2024/european-green-deal_en (Accessed, January 2023).

European Council (2023), Sustainable Finance: Provisional Agreement Reached on European Green Bonds. [online]. Available from: https://www.consilium.europa.eu/en/press/press-releases/2023/02/28/sustainable-finance-provisional-agreement-reached-on-european-green-bonds/ (Accessed, May 2023).

FCA (2022), FCA proposes new rules to tackle greenwashing. [online]. Available from: https://www.fca.org.uk/news/press-releases/fca-proposes-new-rules-tackle-greenwashing (Accessed, May 2023).

FSB (2015), Task Force on Climate-related Financial Disclosures. [online]. Available from: https://www.fsb-tcfd.org/ (Accessed, January 2023).

FSB (2017), Recommendations of the Task Force on Climate-related Financial Disclosures. [online]. Available from: https://www.fsb-tcfd.org/publications/ (Accessed, January 2023).

FT (2022), A whistleblower's greenwashing allegati ons, and the impact they've had. [online]. Available from: https://channels.ft.com/en/ft-moral-money/a-whistleblowers-greenwashing-allegations-and-the-impact-theyve-had/ (Accessed, November 2023).

GFANZ (2022), Enhamcements to Measuring Net-Zero Portfolio Alignment. [online]. Available from: https://www.gfanzero.com/press/gfanz-unveils-enhancements-to-measuring-net-zero-portfolio-alignment-for-financial-institutions/ (Accessed, January 2023).

ICMA (2021), The Green Bond Principles. [online]. Available from: https://www.icmagroup.org/sustainable-finance/the-principles-guidelines-and-handbooks/green-bond-principles-gbp/ (Accessed, May 2023).

Lombard Odier (2021) Designing Temperature Alignment. [online]. Available from: https://am.lombardodier.com/gb/en/contents/news/white-papers/2021/july/designing-temperature-alignment.html (Accessed, January 2023).

MSCI (2022), Implied Temperature Rise. [online]. Available from: https://www.msci.com/our-solutions/climate-investing/implied-temperature-rise (Accessed, January 2023).

PRI (2018), Fiduciary Duty in the 21st Century: France Roadmap. [online]. Available from: https://www.unpri.org/fiduciary-duty/fiduciary-duty-in-the-21st-century-france-roadmap/3843.article (Accessed, May 2023).

PRI (2021), Inevitable Policy Response. [online]. Available from: https://www.unpri.org/sustainability-issues/climate-change/inevitable-policy-response (Accessed, January 2023).

Republique Francaise (2015), La transition energetique pour la croissance verte. [online]. Available from: https://www.legifrance.gouv.fr/jorf/article_jo/JORFARTI000031045547 (Accessed, January 2023).

SEC (2022a), SEC Charges BNY Mellon Investment Adviser for Misstatements and Omissions Concerning ESG Considerations. [online]. Available from: https://www.sec.gov/news/press-release/2022-86 (Accessed, November 2023).

SEC (2022b), SEC Charges Goldman Sachs Asset Management for Failing to Follow its Policies and Procedures Involving ESG Investments. [online]. Available from: https://www.sec.gov/news/press-release/2022-209 (Accessed, November 2023).

TCFD Hub (2021), Measuring Portfolio Alignment. [online]. Available from: https://www.tcfdhub.org/wp-content/uploads/2021/10/PAT_Measuring_Portfolio_Alignment_Technical_Considerations.pdf (Accessed, January 2023).

The Times (2023), The Root of the Problem with Climate Pledges to Plant Trees: There's Not Enough Space. [online]. Available from: https://www.thetimes.co.uk/article/the-root-problem-with-climate-pledges-to-plant-trees-theres-not-eno ugh-space-9n8n9fmb3 (Accessed, January 2023).

TotalEnergies (2020), A Net Zero Company by 2050, Together With Society. [online]. Available from: https://totalenergies.com/company/transforming/amb ition/net-zero-2050 (Accessed, May 2023).

UK Government (2019), UK Becomes First Major Economy to Pass Net Zero Emissions Law. [online]. Available from: https://www.gov.uk/government/news/uk-becomes-first-major-economy-to-pass-net-zero-emissions-law (Accessed, January 2023).

UNFCCC (2015), The Paris Agreement. [online]. Available from: https://unf ccc.int/process-and-meetings/the-paris-agreement/the-paris-agreement (Accessed, January 2023).

WMO (2023), Global Temperatures Set to Reach New Records in Next Five Years. [online]. Available from: https://public.wmo.int/en/media/press-release/glo bal-temperatures-set-reach-new-records-next-five-years (Accessed, June 2023).

10

Other Issues

The To-Do List Grows

In July 2015, I attended a conference titled, "the third international conference on financing for development" in Addis Ababa, Ethiopia (FFD3 for short).

It was my first introduction to the world of development diplomacy. The negotiations themselves were undertaken by heads of states or their representatives. The result was a 37-page negotiated outcome called the "Addis Ababa Action Agenda" (United Nations 2015a).

NGOs with carefully prepared briefing materials worked with a Minister or two, in order to ensure text relevant to the NGO's area of expertise (say, gender equality or children's rights) was included in the negotiated document.

On my part, it was mostly to attend a series of fringe events, to discuss various parts of the financing for development agenda. A few European and US banks and investors attended, although our participation was mostly outside of the official negotiations.

I did participate in one negotiation as part of a small delegation of responsible investment organisations.

The negotiators spoke in a lexicon unfamiliar to me. Particular care is taken to words like "note", "acknowledge", "consider", "recognise", "reaffirm" and "commit", as well as the use of the conditional tense or the future tense. There is a hierarchy here and some sentences would be subject to hours of discussion.

To give an example, paragraph 47 of the Addis Ababa Action Agenda refers to private investment in infrastructure: "We call upon standard-setting bodies

© The Author(s), under exclusive license to Springer Nature
Switzerland AG 2023
W. Martindale, *Responsible Investment*, https://doi.org/10.1007/978-3-031-44536-1_10

to identify adjustments that could encourage long-term investments within a framework of prudent risk-taking and robust risk control." It's very difficult to understand exactly what this would entail. What do we mean by "to identify", what constitutes "adjustments", and what do we mean by "prudent" or "robust"?

I found it overwhelming. I left unsure of the value of my personal contribution and indeed, unsure of the value of the conference.

A few months later, the Addis Ababa Action Agenda contributed to the UN Sustainable Development Goals (SDGs), announced in September 2015 (United Nations 2015b).

The SDGs represent a comprehensive sustainability framework: 17 goals, 169 targets and 232 indicators.

Translating the SDGs into investable themes is complicated. The SDGs were not written for investors, nor even the private sector. The SDGs are the outcome of complicated negotiations, each themselves subject to negotiations, events and outcome documents. It's why the SDGs themselves are very hard to translate into an investment strategy. The choice of indicators and wording around the targets and goals is difficult to understand.

And so, while on the one hand, the SDGs provide investors with an overarching framework to understand sustainable development, there are, inevitably, a number of issues that tend to get more attention from investors than others. In other words, investors have determined a series of more salient sustainability themes that form part of the SDGs.

The dominant issue is climate change, covered in the previous chapter. But there are a number of other issues that are worth exploring.

For most topics, my interpretation of the issue is a mainstream one. Perhaps the exception is modern slavery where, informed by academic and author, Emily Kenway, my interpretation is not typical. I find Kenway's arguments compelling and they've informed mine.

The following sections are a brief overview of some of the more common ESG issues, how I think about them and how they're relevant to investors.

Biodiversity

First, biodiversity.

The other side of the climate change coin is biodiversity loss, a topic that deserves investors' attention in its own right, with many experts considering it a threat more serious than climate change.

A cartoon circulating on NGO websites depicts waves of crises in orders of magnitude. First, Covid 19, then recession, then climate change, and last, the biggest risk of them all, biodiversity collapse. The World Economic Forum seems to agree. Biodiversity loss is rising in its annual ranking of global risks, currently coming in third (World Economic Forum 2023).

But what is biodiversity and how is it relevant to investors?

There are three terms that are often used interchangeably: Biodiversity, nature and natural capital. Nature is the term most familiar to us and can be defined as the natural world untouched by humankind. Biodiversity is the variability of living things. In other words, we can think of nature as land and biodiversity as the richness of that land.

Natural capital is "the stock of resources (e.g. plants, animals, air, water, soils, minerals) that combine to yield a flow of benefits to people" (Natural Capital Coalition 2018). The definition is taken from the Natural Capital Protocol, a group set up in 2016 that works to standardise how we understand these topics.

The economic value of natural capital is considerable. The construction industry, food and beverage sector, agricultural and apparel sectors, chemicals and materials, travel, tourism and real estate are all dependent on natural capital, including issues such as soil quality, access to sources of fresh water, water filtration, pollination and reliable weather patterns. The second-order effects are systemic, including disruption to food supply chains.

In 2016, I joined a group of investors visiting palm oil plantations in Indonesia. Palm oil is an edible vegetable oil. WWF estimates that more than half of all packaged goods Americans consume contain palm oil—it's in lipsticks, soaps, detergents and even ice cream (WWF 2022).

Palm oil is a major cause of deforestation, and can be found in most investors portfolios, along with other commodities that cause deforestation such as beef, soy and timber.

From Jakarta, we flew to Pekanbaru, the economic capital of Sumatra Island. From Pekanbaru we drove about 150 kms south east.

The roads were slow, a mix of tarmac and dirt, and it took just over 4 hours. The clock on the dashboard of the Landcruiser clicked by but the scenery was essentially unchanged. Row after row after row of palm trees standing in regiment like a military parade. There was little in between the palm trees, just sandy strips of dirt, perhaps a fern, perhaps some grass. Each palm was planted with precision, not a metre of space to be wasted.

The lack of biodiversity was by design. Animals could disrupt palm oil yields.

Palm oil is an issue on four fronts. First, the rainforest, which acts as a carbon sink, is bulldozed. Second, any remaining forestry is burnt. The ground is high carbon peatland releasing toxic clouds of GHG emissions into the atmosphere. The fires can continue underground for months. Third, palm trees are planted creating a monoculture, degrading soil.

Fourth, there is a social dimension too. At one of the plantations, the workers wore brand new goggles, yellow vests, sturdy trousers and shoes. But I didn't need to look too far to see barefoot children with a talent for climbing trees. Modern slavery, including child and forced labour, is common.

This is not an issue unique to Indonesia. Many a tropical country is deforesting apace. Indeed, take a train across the British or French countryside and its field after field of wheat, barley, sunflower or grape.

While protecting and restoring biodiversity loss ultimately requires a policy response, there are a few reasons why investors would address biodiversity in their portfolios:

1. Company-specific financial risk. If companies in our portfolios rely on natural capital, nature and biodiversity loss is a threat to companies' profitability. An interesting example is pollination. Supply chains of companies in which we invest may rely on a variety of pollinators to ensure that each plant is pollinated at the right time. There are examples where farmers are having to "install" beehives because there is no longer sufficient natural pollination.
2. Systemic financial risk. Disruption to supply chains could affect multiple parts of our portfolio at the same time in the same direction, for example, a food crisis. An example is the cereal trade. Disruption to the cereal trade would have global impact, particularly in low-income communities. Another example is the steep rise in coffee price due to drought in Brazil. The drought is directly related to deforestation impacting local weather patterns.
3. Reputational risk. NGOs are increasingly engaging companies—and investors—on deforestation. In turn, consumer pressure and changing consumption habits can impact a company's reputation.
4. Engagement opportunity. Deforestation is rising on the agenda of shareholder AGMs.
5. Policy risk. Efforts to stop deforestation involve a complex web of domestic policy, supply-side policies, import policies, international diplomacy and corporate regulation.

Pension schemes can hold their asset managers and advisers to account asking: What are the at-risk sectors and at-risk companies in our portfolios? Which metrics, or combination of metrics, are used to identify risks? How is the asset manager addressing nature and biodiversity-related risks? What are the upcoming shareholder resolutions on deforestation?

There are a number of steps investors can take:

The first is to incorporate PBAF, the Partnership for Biodiversity Accounting Financials, in investment decision-making. PBAF provides guidance on how to understand, assess, measure and take action to mitigate biodiversity-related risks.

The second is TNFD, the nature equivalent to TCFD. TNFD has helped standardise and organise nature-related disclosures across the financial intermediation chain, proposing methodologies to measure nature-related financial risks and support nature-positive investment and publish guidance on target setting (TNFD 2022).

The frameworks are helpful, because measuring nature and biodiversity loss is complex. A ton of greenhouse gas, whether emitted in the UK or Indonesia, has the same atmospheric effect. But an acre of land in the UK will have very different environmental characteristics to Indonesia. On its own, an aggregated portfolio-level metric is not particularly useful, as it depends on sector and geography.

Nick Robins told me, "I'm really surprised by the adoption of nature by the investment community and the wider financial community (such as central banks), because I don't yet see the technological drivers for rescue in the same way as I do for energy transition. From a portfolio view, there are net benefits with energy transition. For nature transition, there are much tougher trade offs."

Groups like Finance for Biodiversity seek to raise awareness, undertake research and pool expertise.

Water

A topic rising on the agenda is water.

A company's freshwater footprint is very hard to measure. Companies use water for a variety of reasons. CDC, the US centre for disease control, cites the US Geological Survey, to explain that "industrial water is used for fabricating, processing, washing, diluting, cooling, or transporting a product" (CDC 2016). This includes the food we eat, the clothes we wear, and the minerals used in our cars, computers and phones. CDC says, "Large amounts of water are used mostly to produce food, paper, and chemicals", with Louisiana topping the list of States' use of freshwater. Conservation Gateway claims that nearly two-thirds of water consumption goes into corporate supply chains.

First, terminology. Water withdrawal is the amount of water a company uses (without considering discharge). Water consumption is water withdrawal minus discharge, assuming the discharge does not contain untreated pollutants. The water stress of a region is the extent to which water is scarce. Finally, water pollution is a measure of a company's waste water discharge.

Some data providers publish waste water discharge, but it is very hard to measure which pollutants and where. Companies normally operate under a permit system, where a company is allowed to pollute in a specific area, but there are few examples where pollution limits are set based on what the environment can handle.

Discharging water in itself is not necessarily an issue. Rather, the polluting substances are an issue, including where the substance is discharged and how soluble it is.

Water neutrality is twofold. First, whether the company extracts less water from a water scarce region than nature can replenish. Second, whether a company pollutes less than nature can refine.

From an investor's perspective, it's then necessary to determine how much of a company's revenue is earned from water stress regions.

Some investors are requiring companies to publish a water policy and to disclose water consumption and pollution. Many companies however say that pollution is within limits set by permits. And that's probably true.

A more sophisticated engagement strategy would also engage local policymakers, and possibly import policies, for agricultural goods and industrial chemicals.

There's a lot more work required for investors to systematically incorporate freshwater issues in portfolio construction, investment, corporate and policy engagement.

The health of oceans and marine life is another issue investors are looking into, perhaps in part, driven by the growth in blue bonds, which the World Bank defines as a "debt instrument issued by governments, development banks or others to raise capital from impact investors to finance marine and ocean-based projects that have positive environmental, economic and climate benefits" (World Bank 2018).

There are a number of environmental issues facing oceans, including overfishing and plastic pollution. Trawlers drag industrial scale nets along the seabed floor releasing carbon-rich sediment into the ocean and atmosphere.

The nets destroy marine life, with considerable bycatch, which is where marine life other than that being fished is caught in the nets. Overfishing is leaving fish stocks severely depleted, with ships fishing where they want due to no or unenforced regulations. Ghost-fishing is where nets are discarded. They simply wash around the oceans for years destroying marine life in their wake.

There are some attempts to establish sustainable fishing but the attempts are wholly inadequate. The problem for investors is that most of the shipping and fishing companies are unlisted. The food companies, or even food retailers, may be a way in for investors, and there are a handful of quality engagements underway, but for most investors, it's unlikely to be a priority for clients, and progress heavily relies on policy change.

Factory Farming and Anti-Microbial Resistance

Jeremy Coller is CIO and chairman of Coller Capital, a private equity fund that specialises in the secondary market; buying and selling existing investments of private companies from other investors.

I forget why we met, but it was in late 2014. A group called The Elders convened a meeting at Coller Capital and Jeremy Coller attended. The Elders was established by Nelson Mandela and included Kofi Annan, the former UN Secretary-General who launched the PRI. Coller is part of The Elders' advisory council and, I believe, a donor.

After the meeting, Coller, a vegetarian, asked me about factory farming. He told me that he'd taken a group of colleagues and clients to an Arsenal football club match and, to his surprise, "wasn't laughed at" when he introduced the idea of a new responsible investment group addressing issues related to factory farming, and in particular, antimicrobial resistance.

FAIRR, the Farm Animal Investment Risk and Return initiative, was launched by Coller in 2015. In 2016, Maria Lettini, a former PRI director, joined as FAIRR's executive director.

FAIRR has grown considerably with, at the time of writing, 38 employees, 15 of which work on collaborative engagements. Through multiple work-streams, the group works on the ESG risks of factory farming as well as the opportunities in sustainable animal agriculture and alternative proteins. The FAIRR investor network now has more than 370 members with $71 trillion in assets around the world.

There are three pillars to FAIRR's work; data and quantitative tools, thematic research and policy guidance, and collaborative engagements. They provide bottom-up analysis on over 120 companies in the food sector that have exposure to animal agriculture.

One of their quantitative tools, the Protein Producer Index, covers protein producers with exposure to the five major categories of farmed animal protein: Beef, dairy (considered separate as the farming processes are quite different), poultry and eggs, aquaculture (which includes salmon, shrimp, tilapia and other fish), and pork. This tool is an ESG benchmark of the 60 largest listed companies involved in the production of animal proteins, across a range of topics such as management of antibiotics, climate change and biodiversity.

Around 70% of global antibiotic use is due to intensively farmed livestock, accelerating the risk of widespread antimicrobial resistance (AMR), a potential systemic risk for investors (FAIRR 2021a). Many antibiotics are shared class, provided to both animals and humans.

From the collaborative engagement perspective, FAIRR is engaging the animal pharmaceutical industry, which is a group of companies that are unlikely to be well-known by consumers but play a significant role in the production of the foods we eat.

The animal pharmaceutical industry is opaque, with few firms disclosing how they manufacture and how they market their products.

The use of antibiotics in animals is not wrong; sick animals should of course be treated, but the problem with intensive factory farming is that antibiotics are given to animals that are not sick for two reasons; to prevent them from getting sick (known as prophylaxis) given poor animal welfare conditions or for growth promotion. Small doses of antibiotics can make animals grow faster.

This is a vicious cycle, because poor animal welfare negates good management of hygiene and ventilation, requiring further use of antibiotics. If antibiotics are overused, bacteria and viruses can create resistance to antibiotics.

FAIRR finds that there is also a growing number of instances of waste, from the production of antibiotics, making its way into water systems. The polluted water streams further exacerbate the AMR risk in the regions of the world where antibiotics are manufactured, mainly China and India.

To combat AMR, the regulatory environment has been strengthened, particularly in Europe, including banning imported meat where antibiotics have been used for animal growth. But as demand for meat grows, so does the use of antibiotics.

In addition to AMR risk, industrial animal farming also drives negative climate and nature impacts. Food production, transportation and waste makes a considerable contribution to GHG emissions. 14.5% of global GHG emissions are linked with animal agriculture, 80% of agricultural land is used for growing and feeding live stock being the leading cause of deforestation and biodiversity loss, also 30% of freshwater is used for animal agriculture.

There are opportunities in terms of more sustainable animal agriculture practices. There are some feed additives that are proven to reduce methane, but doing so is relative, limiting, only to an extent, methane emissions. The bigger prize is to diversify protein mix; how are companies changing their product portfolios to include plant-based proteins, engaging with consumers, product placements and promotions?

As well as supply, investors can engage demand, targeting restaurants to source animal proteins from suppliers that use antibiotics responsible, in line with World Health Organisation (WHO) recommendations.

And an often-overlooked issue with factory farming, which is a leading cause of nutrient pollution, is how to responsibly deal with animal waste, including faeces.

Research by FAIRR, published in 2021, found that the volume of animal faeces produced by the 70 billion land-based livestock processed per year is equivalent to the waste produced by twice the entire global human population, and whereas there are strict regulations for the management of human waste, animal waste remains underregulated (FAIRR 2021b).

If unmanaged, animal waste can lead to environmental degradation and biodiversity loss, polluting soil and waterways. To the contrary, well-managed animal waste can be used as manure and fertiliser in crop production, as well as bioenergy.

Regenerative agriculture is one of the supply-side levers gaining in popularity, which can help agri-food companies meet their climate change commitments, improving soil quality, water quality and filtration benefits, and raising farmer incomes as buyers are comfortable paying premiums, and, for responsible investors, helping investors achieve portfolio sustainability targets.

That said, regenerative agriculture remains a small part of the market and there are discrepancies on definitions, KPIs measure progress and comparability of corporate commitments is missing.

Inequality

Inequality is a topic not often discussed by responsible investors—inequality tends to be personal (how much inequality is too much?), political (how do we address inequality?) and difficult to measure. The relationship between inequality, financial returns and fiduciary responsibilities can be unclear.

When it comes to investor action, there is substantially less political or regulatory clarity than, say, tackling climate change. And there are few off-the-shelf investment products specifically seeking to address inequality.

Inequality is seemingly at odds with the roots of shareholder capitalism. Many companies' business models are based on inequality. Low wage, low skilled, low rights workforces are often a feature of companies' profitability.

Nathan Fabian said to me, "It's harder to get agreement on the necessary actions on issues such as inequality without some benchmark or some objective way to test 'what is good'. In global capital markets, these are the kinds of tools which are easier to respond to."

"For example, on inequality, I think there are benchmarks but they're not necessarily widely agreed or adopted, which makes it harder for investors."

Many savers are asking their pension funds to do more to address social issues.

"One of the few schemes in the UK that does elicit members' views is the railways pensions scheme. Its members are pro-responsible investment and amongst ESG issues, they put treatment of workers first and climate second", ShareAction's CEO, Catherine Howarth, told me.

"There's no doubt in my mind that a more member-centric investment strategy would rebalance towards the S of ESG."

"I'm sure that climate and biodiversity would continue to have prominence, not least because of the financial materiality of these issues, but there would be more attention overall to social issues."

Philippe Zaouati, founder and CEO at Mirova, agrees. "To be frank, my background is much more on social issues."

"I didn't come from a well-off family, I went to a school in Marseille that was ranked as one of the worst 1% schools. So that informs my view and commitment to equality of opportunity."

And for David Blood, it was also social issues that led to his passion for responsible investment.

"It goes back to my childhood in Brazil, where I saw very deep poverty. I didn't understand why I lived one way and why children my own age lived in another way. I asked my parents to explain, but they couldn't. My interest in social justice goes back to when I was 11 years old."

"I first met Al Gore in Boston. My interests were in poverty and sustainable development and Al's were in the environment and climate change, and it quickly became clear to us that they were two sides of the same coin: people and planet."

Inequality is also one of the topics I feel most passionately about. It's where we really test what we mean by responsible investment. Inequality is, in large part, the root of all other sustainability issues.

Inequality is linked to most if not all the SDGs, in particular SDG 1 "no poverty", SDG 2 "zero hunger", SDG 4 "quality education", SDG 5 "gender equality" and SDG 8 "decent work and economic growth". SDG 10 "reduced inequalities within and among countries" specifically addresses inequality.

At its heart, the climate crisis is an inequality crisis. While those in Florida or the Netherlands may benefit from flood defences, those in Bangladesh do not.

And in recent economic history, the financial crisis, the Eurozone crisis and the pandemic have made inequality worse.

While more challenging than say, climate change, some investors have attempted to frame inequality as a financial issue, putting forward the financial arguments in favour of tackling inequality.

In the UK, we've seen stubborn wage growth. Whereas the top 1% have seen their share of national income rise, the bottom 50% have seen it fall (The Equality Trust 2020). The Resolution Foundation estimates that almost a quarter of all household wealth in the UK is held by the richest 1% of the population (Resolution Foundation 2020).

We've also seen a rise of "in work" poverty, where, for a growing number of working people, work does not provide a route out of poor living standards. The Joseph Rowntree Foundation found that, in the UK in 2018, 56% of people living in poverty were in a household where at least one person had a job, versus 39% 20 years earlier (Joseph Rowntree Foundation 2020).

These issues are not unique to the UK.

A society, where, due to inequality, talented individuals are unable to fulfil their economic potential, has implications for GDP and therefore for growth and international competitiveness.

Many attribute the rise in populism across the world to heightened inequality, with populist politics playing to the disenchantment of those left behind by offering simple vote winning solutions to complex problems. Populism can be unpredictable, lead to divestment, both within countries and from overseas investors.

The response from financial policymakers on social issues is mixed. Some policymakers have begun to turn their attention to social issues, but they lag by some way environmental issues.

In 2021, the UK DWP consulted on social risks and opportunities, but rather than regulate disclosure (as is the case for climate change), it formed a social issues taskforce (DWP 2022). In the EU, the technical expert group has started to consider social issues, but several years in arrears to climate change, and other environmental issues.

We have seen some attention to social issues through corporate engagement—including on executive pay, tax fairness and labour rights. But there's nothing like the frameworks we see in place on climate change.

Although less well-developed, there are a series of actions investors can take. For example, direct investments in social housing, social infrastructure, as well as emerging themes such as rehabilitation bonds. I'd like to see a sovereign issue a SDG or social issues bond to help grow the market (Uruguay's sustainability-linked bond provides us with a good case study). Investors are beginning to engage companies on their tax arrangements, approach to freedom of association and living wages.

That said, I think inequality is a big challenge for investors. Investors may be able to do some things to help address inequality at the margins. But at a system level it does not appear that investors have the tools or incentives to address the causes of inequality in a way that would affect system-level risks.

And it's difficult to seriously tackle inequality, without thinking about growth.

I asked Erinch Sahan, business and enterprise lead at the Doughnut Economics Action Lab about his priority sustainability issue.

"The issue that most drives my work is inequality, including income inequality, but particularly wealth inequality."

"Inequality is an ecological issue. If there is a limited ecological footprint that our economic activity can have, then the way we distribute the benefits of that economic activity is critical."

"Inequality is a good indicator for a broad range of social impacts as well. If we're increasingly channeling value and opportunity into the hands of fewer and fewer people, we will undermine our ability to meet the needs of all within the means of the living planet."

"When it comes to the social and ecological, we cannot decouple the two."

"If we rely on unending economic growth to trickle down enough 'stuff' for everybody, we'll break planetary boundaries."

On GDP, Fiona Reynolds agreed. "When we think about how we measure the success of the economy, we measure GDP, which looks at what we consume. We don't look at the outcomes of that consumption. That's a system failure that needs to change."

Further work on inequality should—and I expect, will—be part of responsible investment's future, both in terms of actions investors should take to address inequality, as well as more frank conversations about responsible investment's limitations.

Human Rights

The Oxfam campaign, Behind the Brands, was my first introduction to human rights frameworks.

Behind the Brands assessed six (later extended to 10) food and beverage companies across a range of sustainability themes. The companies were well-known brands: Nestle, Danone, Unilever, Associated British Foods, Kellogg's and Mars. Most companies, like Unilever, are listed, some were not, like Mars.

A team of researchers downloaded each company's public policies and carefully reviewed those policies against a series of indicators determined by thematic experts. My focus was on Free Prior and Informed Consent (FPIC).

FPIC allows communities to withhold consent for a project. In the case of Behind the Brands, the project was often for farming, or the construction of a mill, on land that may have historical or religious meaning. FPIC can be negotiated and subject to certain conditions. Companies shouldn't be discouraged from proposing a project, as long as the project is designed, implemented, monitored and evaluated consistent with conditions established by the UN, NGOs, local government and of course local people.

I reviewed company websites, reading and re-reading policies to understand whether FPIC was universally applied, the company's approach to negotiation and remedy.

Once Behind the Brands was published my job was to take the analysis to investors and banks. I was interested in understanding how they considered human rights issues.

One meeting was with Lord Deben, later appointed chair of the UK climate change committee.

As farcical as it sounds, we met for breakfast, eating egg white omelettes cooked by, I presume, his chef, while discussing our findings. Lord Deben was an adviser to one of the companies.

He was engaged, thoughtful and agreed to take the findings forward, including in conversations with investors.

Another meeting was with an equity analyst at Morgan Stanley. The research was useful, I was told, and it would be incorporated in analysis Morgan Stanley would publish for their investor clients.

But in most cases, I didn't make much progress. In 2013, human rights were not well understood by investors.

For those who are not human rights experts, John Ruggie's "Just Business" is a must read (Ruggie 2013). Part biographical, part analysis, it is the story of how Ruggie put human rights firmly on the agenda of companies and investors.

The UN defines human rights as rights we have simply because we exist as human beings—they are not granted by any state. "These universal rights are inherent to us all, regardless of nationality, sex, national or ethnic origin, color, religion, language, or any other status. They range from the most fundamental - the right to life - to those that make life worth living, such as the rights to food, education, work, health, and liberty" (United Nations 1948).

In 1948, the UN General Assembly adopted the Universal Declaration of Human Rights. However, the idea that business enterprises "might have human rights responsibilities independent of legal requirements in their countries of operation is relatively new and still not universally accepted" (Ruggie 2013).

Ruggie says, "Human rights traditionally have been conceived as a set of norms and practices to protect individuals from threats by the state, attributing to the state the duty to secure the conditions necessary for people to live a life of dignity."

The result, after years of shoestring budgets, research by a group of overworked analysts and expert diplomacy, was the UN Guiding Principles on Business and Human Rights (OHCHR 2011). Testament to Ruggie's perseverance, the GPs were unanimously endorsed by the member states of the UN human rights council.

The GPs, says Anita Dorett, Program Director at the Investor Alliance for Human Rights, "is the authoritative global framework for addressing business impacts on human rights" (Business and Human Rights resource Centre 2021).

The GPs help establish the role of state and companies, on the impact of companies' economic activities on human rights.

There are three pillars, first, the role of the state to protect against human rights abuse, second, the corporate responsibility to respect human rights, to avoid human rights abuse, and to address adverse human rights impacts, and third, remedy, such that those who are victims of human rights abuses caused by company activity can seek appropriate compensation.

The GPs are now widely embedded in domestic and international law and standard setters' frameworks. "Even by FIFA, the international football association, in the bidding to host the world cup" (Ruggie 2013).

The GPs include a framework to understand and address human rights risks. The GPs were followed by a 2017 publication by the Organisation for Economic Co-operation and Development (OECD).

The OECD is perhaps one of the strangest international organisations I've worked with. Headquartered in Paris, the OECD is funded by its member states dependent on the size of economy. As such, the US is its largest contributor. Under the Trump administration, the OECD, which is effectively a series of secretariats covering a range of policy topics, toed a careful line on sustainable finance topics, not wanting to attract unwelcome attention from US funders.

Its building is a maze of underground corridors with high security, reflecting both the OECD's importance and perhaps the heightened state of terrorist threat in Paris.

The responsible business conduct (RBC) for institutional investors (published in 2017) followed the multinational enterprise (MNE) guidelines (adopted in 1976, and revised multiple times since). The MNE guidelines established national contact points (NCPs), a platform whereby victims of human rights abuse can seek remedy. Most experts say that the NCPs are not well enforced. Companies would however want to avoid being subject to NCP investigation.

Responsible business conduct sets out expectations for investors undertaking due diligence in new and existing investment activity. The 66-page document includes a, now dated, overview of responsible investment, fiduciary duties, ESG integration and active ownership (OECD 2017).

More importantly, it introduced a three-part framework for addressing adverse impacts depending on whether the investor caused, contributed to, or was directly linked to human rights abuse.

In the case of caused, the investor is responsible for remedy. In the case of contributed to, it must cease its investment, and use leverage to mitigate adverse impacts. If directly linked, it must use leverage to mitigate adverse impacts.

Most investment comes under "directly linked". The complexity of the intermediation chain detaches investor from company and company from supply chain.

A large multinational food company, for example, that provides tinned fish to UK supermarkets sources the fish from a series of private companies operating in the seas around Malaysia.

One of the suppliers to that private company is a fishing vessel with enslaved fishermen that have migrated from Myanmar. The fishing vessel forces these men, who are undocumented and escaping conflict, to fish for months on end, with no or little pay, no contact with family, and regular use of violence and intimidation.

To what extent is the investor responsible for the human rights abuse? RBC says they are directly linked and should use their leverage "to influence the entity causing the adverse impact to prevent or mitigate the impact". That may include for example, joining a collaborative engagement to address the multinational food company's sourcing policies.

While inadequate, due diligence across all portfolio holdings is disproportionately expensive and it's unclear where exactly responsibility ends.

Richard Roberts, Inquiry Lead at Volans, said to me, "We should all try to act in a way that is morally justifiable. If you've identified there is a potential human rights issue, your moral compass should be strong enough to say, we should be seriously engaging on the topic, even if the issue is not financially material."

The UN Global Compact preempted the Guiding Principles. Global Compact, or UNGC, has 10 principles (United Nations 2000). Principles 1 and 2 are as follows:

"Principle 1: Businesses should support and respect the protection of internationally proclaimed human rights; and

Principle 2: make sure that they are not complicit in human rights abuses."

It's a voluntary initiative. Companies sign up via CEO statement and pay a (small) fee. 13,000 companies subscribe. UNGC has a series of local networks, which vary in quality. UNGC convenes companies and publishes research reports, but there is little monitoring or enforcement. There are attempts by ratings providers to determine company's compliance. MSCI's indicator is described as, "whether the company is in compliance with the United Nations Global Compact principles", with a more detailed methodology on ESG controversies and global norms.

Some investors will prioritise their engagement or exclude companies based on their, or a third-party's, assessment of the company's activities against the UNGC's principles.

In 2020, PRI published a paper titled, "why and how investors should act on human rights", essentially making the case for further investor action (PRI 2020). It's one of PRI's stronger publications, both explaining human rights and the role of investors, but also committing PRI to further action. "We are therefore setting out a multi-year agenda for our work towards respect for human rights being implemented in the financial system", PRI says.

In 2022, PRI launched ADVANCE, the human rights equivalent of Climate Action 100+. It seeks to support investor collaborative engagement in companies most at risk from human rights abuses. Based on the levels of investor interest, it's a collaboration that I expect will have impact.

We've also started to see the introduction of due diligence requirements on human rights, in France, the UK and in Europe, with the introduction of the Corporate Sustainability Due Diligence directive.

I asked Eelco van der Enden, CEO at GRI, about investor due diligence requirements, "You're seeing due diligence in multiple fields, not just sustainability, where the burden of due diligence is being shifted to the investor."

"You see it, for example, with banks when it comes to money laundering. It is the role of banks to undertake due diligence. Or, for example, with companies, when it comes to tax. It is the role of companies to file their tax return."

"This is a global trend whereby the burden of due diligence no longer lies with the supervisory organization, but is transferred to the regulated entity."

"I don't think I'm particularly in favour. There is a fiduciary duty of investors to make sure that what you're investing in is right for the end saver. But if it would go beyond a reasonable level of due diligence then I would say that it's a step too far."

While responsible investors are paying more attention to human rights issues, many consider more concerted action the remit of policymakers, not investors.

Modern Slavery

Companies tend to consider the salience of human rights issues when assessing risks in their supply chain: "A company's salient human rights issues are those human rights that stand out because they are at risk of the most severe negative impact through the company's activities or business relationships" (UN Guiding Principles 2016).

Perhaps one of the most salient human rights risks for companies and investors is "modern slavery".

While I was at PRI, PRI's CEO, Fiona Reynolds, chaired the Liechtenstein Initiative's Financial Sector Commission on Modern Slavery and Human Trafficking.

The International Labor Organisation (ILO) defines modern slavery as follows: "Essentially, it refers to situations of exploitation that a person cannot refuse or leave because of threats, violence, coercion, deception, and/or abuse of power" (ILO 2017). In the UK, addressing modern slavery was a major part of Theresa May's short tenure as prime minister, describing modern slavery as the "great human rights issue of our time".

The 2015 Modern Slavery Act requires companies to produce an annual slavery and human trafficking statement on their website's homepage.

I asked Fiona Reynolds, PRI's former CEO, about her focus on modern slavery. "When I came to the PRI, I never felt that there was enough attention on social issues from the signatory base or the PRI itself."

"While the dial has moved, I still don't think there is enough attention on social issues. I think we still deny the scale of modern slavery and human trafficking. It is an international crisis and many of the companies we invest in and buy from exploit people for profit, sometime sadly knowingly and sometimes from a lack of due diligence on their supply chains."

"Reporting on Modern Slavery in countries like the UK, France and Australia has put a focus on modern slavery and human trafficking. More people are aware of these issues, but they do not act to the extent we need to."

"I hope that some form of mandatory human rights due diligence will be part of the answer. The issues need raising in the same way that we have elevated climate change."

But in truth, I didn't understand modern slavery, nor the role of investors, until reading Emily Kenway's The Truth About Modern Slavery. Kenway is a campaigner and academic, that worked for the UK's first anti-slavery commissioner. Kenway concludes, "I'm about to make a statement that'll be considered unacceptable by many, but it's necessary: 'rescue' [from modern slavery] is not the best option for some people when there are no better futures for them than the exploitative conditions … This is the reality beneath the modern slavery story" (Kenway 2021).

Given the attention on modern slavery, this is a statement that goes against the grain. Modern slavery, Kenway argues, is a function of economic inequality. "In this hugely unequal global context, with brands driving down prices and accelerating order time frames, it's logical that workers will be exploited and that some of that exploitation will be extreme." Kenway explains "it's purposeful business strategy to locate production in countries with low worker protections where exploitation is more likely to happen."

In my own corporate engagement activities, I find myself faced with well-known brands, like Amazon or Tesla, that refuse to recognise trade unions or push back against raises to minimum wages. "The system that produces precarious workers, a lack of rights protections, relative impunity for employers and brands and so on, is the same system that produces the really awful cases our headlines report as slavery," says Kenway. "Modern slavery is not a separate phenomenon from general working conditions."

"If we are talking about legal ownership of another human being … then yes, slavery by large was ended in the nineteenth century. But if we are talking about exploitation in terms of being 'modern slavery', there is nothing modern about it … it is the continuation of exploitation by subtler means …" It has "never been something with which nation-states have a problem; rather, it has been something they have created and perpetuated actively because it brings profit."

If shareholder capitalism, and the economic inequality it causes, is the root of what we call modern slavery, what, if anything, should shareholder capitalists do?

Kenway would do away with the modern slavery framing—not to say that slavery-like conditions do not exist, but rather that the modern slavery framing, with policymakers, investors, companies and NGOs as the "great liberators" is not just unhelpful, but counterproductive. Kenway's arguments are compelling and at their heart is system change, ending her book with, "In doing so, we would lose our fairy tales, but we would gain a better reality."

To accept Kenway's arguments would be for investors to redirect their attention to issues such as inequality, erosion of social safety nets or redistributive tax policy. This is not the path investors have followed.

I think Kenway helps put modern slavery in context, a symptom, not a cause.

The ILO and Walk Free estimate that 49.6 million people are in modern slavery. It's an extraordinary statistic (ILO 2021). That's one in every 152 people. It's impossible to divest from modern slavery, it's too prevalent.

There is increasing regulation too. While regulation tends to require disclosure, for repeat offenders, this may lead to a ban.

Investors would tend to engage companies assessing the likelihood of modern slavery based on the vulnerability of workforce, supply chain geography, and in particular, exposure to conflict or mass migration, products and business models. The more proficient investors would engage policymakers too.

But the more effective action, in my view, would be to address the causes of modern slavery—perhaps starting by thinking about how investors can address inequality.

Gender Inequality

While gender inequality is, I think, well understood by investors, its interpretation is broad.

Board diversity is a well-established engagement theme, with a gradation of expectations depending on the market. For example, in Japan or the Middle East, board diversity can be more challenging than Canada or the UK, but regardless of the country, most investors will challenge companies with all male boards.

Pay gaps are also a well-established theme, both horizontal pay gaps where women are paid less than men for the same responsibilities, as well as vertical pay gaps where firms are skewed towards male management. The same is true in industry, where more physical work, such as working at warehouses, that tend to favour men, are paid more than roles that tend to favour women.

In a step I thought particularly innovative, one investor compared management reporting to LinkedIn updates, cross-checking that companies that say they have put in place processes to promote women to senior roles is evidenced by their employees' LinkedIn updates.

Investors also engage on discrimination, ensuring companies have whistle-blowing policies in place and management takes steps to enable cultures that allow women that are victims of discrimination to come forward, and are supported in doing so.

More challenging issues are supply chain policies that go beyond legal requirements, such as maternity and paternity pay, flexible working to support mothers and fathers with childcare responsibilities, support for women through menopause, access to sexual health, in countries or US states that restrict access, access to safe routes for abortion, as well as illegal issues, but too often unenforced, such as zero tolerance for violence against women and girls, including trafficking.

Initiatives such as the 30% club, the diversity project, and NGOs such as Oxfam or human rights groups are taking steps to elevate gender equality as an investment theme.

The UN defines gender equality as: "The equal rights, responsibilities and opportunities of women and men and girls and boys. Equality does not mean that women and men will become the same but that women's and men's rights, responsibilities and opportunities will not depend on whether they are born male or female."

"Equality between women and men is seen both as a human rights issue and as a precondition for, and indicator of, sustainable people-centered development" (UN Women 2001).

UN SDG 5 is: "Achieve gender equality and empower all women and girls." (United Nations 2015b). The UN says: "Gender equality is not only a fundamental human right, but a necessary foundation for a peaceful, prosperous and sustainable world. There has been progress over the last decades, but the world is not on track to achieve gender equality by 2030."

The World Benchmarking Alliance publishes an assessment of companies' policies and practices on gender equality across governance and strategy, representation, compensation and benefits, health and well-being and violence and harassment. This includes assessments of gender equality in leadership, gender pay gap and living wages, paid carer leave, safe and healthy working environment, violence and harassment prevention, grievance mechanisms, and training and support.

Gender equality is a moral, but also an economic opportunity. Increased women's labour participation contributes to growth, competitiveness and productive capacity. The UK gender pay gap is 8.3% (UK Parliament 2022). Women in senior leadership may challenge homogenous decision-making structures. Yet, just 9% of Fortune 500 companies are led by women (44 of 500) (Fortune 2022).

New jobs occupied by women are more likely to reduce poverty (EIGE 2016). Women have less access to financing to start a business, are subject to gender-biased credit scoring and gender stereotyping in investment valuations, and more reliant on self-financing. Women are more likely to be victims of discrimination, harassment and sexual violence (OECD 2015).

There is a lot more for investors to do to tackle gender inequality.

References

Business and Human Rights Resource Centre (2021), Investor Responsibility Under the UNGPs. [online]. Available from: https://thiswayup.sounder.fm/episode/investor-responsibility-un-guiding-principles-ungp (Accessed, January 2023).

CDC (2016), Industrial Water. [online]. Available from: https://www.cdc.gov/healthywater/other/industrial/index.html (Accessed, January 2023).

DWP (2022), New Taskforce to Support Pension Scheme Engagement with Social Factors in ESG Investing. [online]. Available from: https://www.gov.uk/govern

ment/news/new-taskforce-to-support-pension-scheme-engagement-with-social-factors-in-esg-investing (Accessed, May 2023).

Edmans, Alex (2011), Does the Stock Market Fully Value Intangibles? Employee Satisfaction and Equity Prices. [online]. Available from: http://faculty.london.edu/aedmans/Rowe.pdf (Accessed, June 2023).

Edmans, Alex (2023), Diversity, Equity, and Inclusion. [online]. Available from: https://papers.ssrn.com/sol3/papers.cfm?abstract_id=4426488 (Accessed, June 2023).

EIGE (2016), Economic Benefits of Gender Equality. [online]. Available from: https://eige.europa.eu/gender-mainstreaming/policy-areas/economic-and-financial-affairs/economic-benefits-gender-equality (Accessed, January 2023).

Kenway, Emily (2021), The Truth About Modern Slavery.

FAIRR (2021a), Animal Pharma. [online]. Available from: https://www.fairr.org/research/animal-pharma/ (Accessed, June 2023).

FAIRR (2021b), How Animal Waste Mismanagement Drives Biodiversity Loss and Accelerates Climate Risk. [online]. Available from: https://www.fairr.org/index/spotlight/animal-waste-spotlight/ (Accessed, June 2023).

ILO (2017), Global Estimates of Modern Slavery: Forced Labour and Forced Marriage. [online]. Available from: https://www.ilo.org/global/publications/books/WCMS_575479/lang--en/index.htm (Accessed, January 2023).

The Equality Trust (2020), The Scale of Economic Inequality in the UK. [online]. Available from: https://equalitytrust.org.uk/scale-economic-inequality-uk (Accessed, January 2023).

Fortune (2022), The Number of Women Running Fortune 500 Companies Reaches a Record High. [online]. Available from: https://fortune.com/2022/05/23/female-ceos-fortune-500-2022-women-record-high-karen-lynch-sarah-nash/ (Accessed, January 2023).

ILO (2021), Frequency Asked Questions on Just Transition. [online]. Available from: https://www.ilo.org/global/topics/green-jobs/WCMS_824102/lang--en/index.htm (Accessed, May 2023).

Ruggie, John (2013), Just Business: Multinational Corporations and Human Rights.

Joseph Rowntree Foundation (2020), What Has Driven the Rise of In-Work Poverty? [online]. Available from: https://www.jrf.org.uk/report/what-has-driven-rise-work-poverty (Accessed, January 2023).

https://www.worldbank.org/en/news/feature/2018/10/29/sovereign-blue-bond-issuance-frequently-asked-questions.

Living Wages Foundation (2022), What Is the Living Wage? [online]. Available from: https://www.livingwage.org.uk/ (Accessed, January 2023).

Natural Capital Coalition (2018), Natural Capital Protocol. [online]. Available from: https://naturalcapitalcoalition.org/wp-content/uploads/2018/05/NCC_Protocol_WEB_2016-07-12-1.pdf (Accessed, January 2023).

OCHCR (2011), Guiding Principles on Business and Human Rights. [online]. Available from: https://www.ohchr.org/sites/default/files/Documents/Publications/GuidingPrinciplesBusinessHR_EN.pdf (Accessed, January 2023).

OECD (2015), Do Women Have Equal Access to Finance for Their Business. [online]. Available from: https://www.oecd.org/gender/data/do-women-have-equal-access-to-finance-for-their-business.htm (Accessed, January 2023).

OECD (2017), Responsible Business Conduct for Institutional Investors. [online]. Available from: https://mneguidelines.oecd.org/RBC-for-Institutional-Investors.pdf (Accessed, January 2023).

PRI (2020), Why and How Investors Should Act on Human Rights. [online]. Available from: https://www.unpri.org/human-rights/why-and-how-investors-should-act-on-human-rights/6636.article (Accessed, January 2023).

Resolution Foundation (2020), The UK's Wealth Distribution and Characteristics of High-Wealth Households. [online]. Available from: https://www.resolutionfoundation.org/app/uploads/2020/12/The-UKs-wealth-distribution.pdf (Accessed, January 2023).

TNFD (2022), v.03 of the TNFD Beta Framework. [online]. Available from: https://framework.tnfd.global/executive-summary/v03-beta-release/ (Accessed, January 2023).

UK Parliament (2022), The Gender Pay Gap. [online]. Available from: https://commonslibrary.parliament.uk/research-briefings/sn07068/ (Accessed, January 2023).

United Nations (1948), Universal Declaration of Human Rights. [online]. Available from: https://www.un.org/en/about-us/universal-declaration-of-human-rights (Accessed, January 2023).

United Nations (2000), The Ten Principles of the UN Global Compact. [online]. Available from: https://www.unglobalcompact.org/what-is-gc/mission/principles (Accessed, January 2023).

United Nations (2001), Concepts and Definitions. [online]. Available from: https://www.un.org/womenwatch/osagi/conceptsandefinitions.htm (Accessed, January 2023).

United Nations (2015a), Resolution Adopted by the General Assembly on 27 July 2015. [online]. Available from: https://documents-dds-ny.un.org/doc/UNDOC/GEN/N15/232/22/PDF/N1523222.pdf (Accessed, January 2023).

United Nations (2015b), The 17 Goals. [online]. Available from: https://sdgs.un.org/goals (Accessed, January 2023).

United Nations (2015c), Paris Agreement. [online]. Available from: https://unfccc.int/sites/default/files/english_paris_agreement.pdf (Accessed, May 2023).

UN Guiding Principles (2016), Salient Human Rights Issues. [online]. Available from: https://www.ungpreporting.org/resources/salient-human-rights-issues/ (Accessed, January 2023).

World Bank (2018), Sovereign Blue Bond Issuance: Frequently Asked Questions. [online]. Available from: https://www.worldbank.org/en/news/feature/2018/10/29/sovereign-blue-bond-issuance-frequently-asked-questions (Accessed, January 2023).

World Economic Forum (2023), Global Risks Report 2023. [online]. Available from: https://www.weforum.org/reports/global-risks-report-2023/ (Accessed,

January 2023).

WWF (2022), 8 Things to Know About Palm Oil. [online]. Available from: https://www.wwf.org.uk/updates/8-things-know-about-palm-oil (Accessed, January 2023).

11

The EU Leads: The UK's Not Far Behind

Sustainable Finance Action Plan

Martin Spolc, Head of the Unit for Sustainable Finance at the European Commission, said, "I often joke that, while I've been working on sustainable finance for five years now, I've got 10 years older. There are more important people working on sustainable finance than me, but I do think that the Commission and my team have made a very important contribution to this file."

The history is something like this. The Capital Markets Union (CMU), the centrepiece of the Juncker-administration was launched in 2014. In 2015, Juncker published an energy union.

Running alongside the CMU is sustainability. The SDGs launched in 2015. The Paris Climate Agreement agreed in 2015. The European Commission started to understand sustainability as an imperative, but also as a competitive advantage, particularly, in light of the Brexit referendum and election of President Trump, both in 2016. As such, a sustainable capital markets union.

The CMU includes standardised disclosure for pensions, and increasingly, standardised pensions, standardised disclosure for investments, with an EU supervisory framework, and a series of disclosure requirements for investors across asset class (European Commission 2015).

In December 2016, the European Commission launched the High Level Expert Group (HLEG). The HLEG's final report followed in January 2018.

The European Commission's Action Plan: Financing Sustainable Growth followed in March 2018.

© The Author(s), under exclusive license to Springer Nature
Switzerland AG 2023
W. Martindale, *Responsible Investment*, https://doi.org/10.1007/978-3-031-44536-1_11

The introduction to the Action Plan says: "Reorienting private capital to more sustainable investments requires a comprehensive shift in how the financial system works. This is necessary if the EU is to develop more sustainable economic growth, ensure the stability of the financial system and foster more transparency and long-termism in the economy. Such thinking is also at the core of the EU's Capital Markets Union (CMU) project" (European Commission 2018).

The Action Plan has "two urgent imperatives".

1. Improving the contribution of finance to sustainable and inclusive growth by funding society's long-term needs.
2. Strengthening financial stability by incorporating ESG issues into investment decision-making.

For responsible investment, the differentiation was significant. While the two issues are connected, the European Commission frames "ESG integration" as strengthening financial stability, while "sustainable finance" is a contribution to society's long-term needs.

The Action Plan is worth reading, and consists of 10 actions.

First, a Taxonomy, second, green labels, third, what it calls, "fostering investment in sustainable projects". It's not clear to me what this means, nor that the Commission has focused much effort here.

Fourth, incorporating sustainability in financial advice. In short, professional investment advisors must ask clients their sustainability preferences, and make recommendations accordingly. Fifth, sustainability benchmarks. Sixth, ESG ratings. Seventh—the recommendation most personal to me and the work of Fiduciary Duty in the 21st Century—clarifying fiduciary duties.

The word "clarifying" is also significant. The Commission (and the PRI's Fiduciary Duty in the 21st Century) was always clear that duties already required consideration of sustainability issues. Nevertheless, "explicitly requiring asset managers to integrate sustainability considerations" was very welcome.

Eighth, incorporating sustainability into prudential requirements, including capital requirements. Ninth, a range of proposals to enhance sustainability disclosures. And finally, corporate governance and short-termism.

The progress that we've seen since demonstrates just how significant a milestone the Action Plan was.

It's often said that the EU's biggest export is legislation, such as Europe's carbon border adjustment mechanism and emissions trading scheme, as well as sustainable finance regulation, with more and more companies headquartered outside the EU have been brought into the EU's legislative regime.

"Sustainable finance can have enormous influence" Martin Spolc, head of unit for sustainable finance at the European Commission, told me.

> "To meet our net zero goals, investments of more than 700 billion Euros are needed every year, in addition to what has been set out to meet the goals of the European Green Deal, RepowerEU and the Net Zero Industrial Act. We can't do that without the private sector."
>
> "Public money is very important, but the success of achieving our sustainability goals will depend on investors and banks."
>
> "We need to ensure public and private money reinforce each other to meet out sustainability goals. Private capital will be by far the most decisive element, so investors and banks have a really crucial role to play here."

The tri-party relationship between US, China and Europe applies to many policy topics, sustainable finance included. When I met with EU policymakers through the course of 2022 and 2023, I was struck by how peripheral UK policymaking was to EU policymakers.

Europe does not have the materials for the transition, so it requires European policymakers to work with jurisdictions elsewhere, and China and US are the power-brokers. And Europe does not have the public financing, nor political consensus, to rival the Inflation Reduction Act, which in part, explains its focus on sustainable finance.

Member state politicians have also since called for a pause on sustainable finance regulation, most notably French President Emmanuel Macron: "I call for a European regulatory pause [on industrial and environmental affairs]. Now we should be implementing them, not making new changes in the rules or we are going to lose all [industrial] players" (Le Monde 2023).

But while there may be a pause in new regulations, there will not be a pause in the implementation of existing regulations.

The EU Taxonomy

If TCFD is greening finance (by seeking to understand the financial risks and opportunities associated with climate change), taxonomies are financing green (by seeking to understand the extent to which investments are aligned with environmental objectives).

There are several taxonomies in development, but it's the EU Taxonomy that is the most comprehensive.

The EU Taxonomy was the flagship recommendation of the European Commission's Action Plan for Financing Sustainable Growth.

The Taxonomy regulation is in two parts: political and technical (European Union 2021). The political part is, what's called, Level 1 regulation (determined by the political process of European Commission, Parliament and Council). It sets out who must disclose, when and how. Controversially, it includes natural gas and nuclear (under certain conditions) as qualifying as green.

The technical part was overseen by a technical expert group (TEG), chaired by Nathan Fabian, who, at the time, was also PRI's Chief Responsible Investment Officer. The group included investors, NGOs and academics. The TEG was tasked with determining what's green and what's not.

The collaboration between policymaker and technical expert is another interesting feature of the EU Taxonomy process. It's a collaboration that has worked.

The Taxonomy sets out green economic activities, not companies. It uses what are called NACE codes, which is a classification system specific to Europe to categorise economic activities. NACE is a French term, which stands for "Nomenclature statistique des Activités économiques dans la Communauté Européenne".

The US equivalent is GICS (Global Industry Classification System). There is a NACE GICS mapping.

The TEG, with additional expert input, determined relevant economic activities and performance thresholds, for example, the levels of GHG emissions per unit of production that is consistent with the Paris Climate Agreement, for the activities that contribute most to climate change.

There are three types of activities that the EU Taxonomy classifies as green.

1. Activities that are already consistent with the Paris Climate Agreement, for example, solar and wind energy production.
2. Activities that reflect best practice, but are not yet consistent with the Paris Climate Agreement, for example, low carbon cement. As new technologies evolve, the performance thresholds for these activities will be tightened.
3. Activities that contribute to either of the above two activities, such as component parts for electric vehicles.

Investors assess the activities undertaken by the companies in their portfolios, determine whether the activities meet the performance thresholds, and then disclose the total as a percentage of the portfolio.

Following the EU Taxonomy's publication, lots of commentators called for 'shades of green' or in other words to have several performance thresholds.

This is because the percentage of Taxonomy alignment in a typical portfolio is very low, say 10–15%, even for portfolios designated as green, and lower than 5% for traditional portfolios. Shades of green, critics argued, would allow an investor to disclose activities that are making progress towards best practice. The TEG disagreed.

The EU Taxonomy includes six environmental themes. They are climate mitigation, climate adaptation, water, circular economy, pollution prevention and biodiversity. Policymakers have also committed to a social taxonomy, development of which is now underway.

The EU Taxonomy includes do no significant harm (DNSH) and social safeguard requirements. An activity cannot be environmentally sustainable if it does harm to another environmental theme (a windfarm on a floodplain) or it fails to meet minimum social safeguards (components for solar panels made with child labour).

Funds that include some form of environmental objective must disclose against the taxonomy.

The EU Taxonomy is not without its critics, on both sides; those claiming it goes too far, and those claiming it does not go far enough.

In September 2022, a number of NGOs resigned from the TEG in response to the decision by European policymakers to classify gas and nuclear (in certain instances) as green.

When I spoke with a member of staff at the European Commission about the inclusion of gas and nuclear, the response was "whatever your view on the inclusion of gas and nuclear, this is still the most advanced regulatory tool out there."

The NGOs did not resign at the time when policymakers decided to categorise gas and nuclear as green but several months later. One commentator pinned the blame on Pascal Canfin, a French MEP that used to head WWF, as having instigated the inclusion of gas and nuclear. Another told me that the NGOs' involvement in the TEG would not be renewed. In other words, they jumped before they were pushed.

The inclusion of gas and nuclear in definitions of green was a setback for the Taxonomy because it undermined the independence of the TEG.

I asked Nathan Fabian, the then chair of the TEG, about the inclusion of gas and nuclear. "I think the primary driver for the EU's inclusion of gas and nuclear was the different priorities around energy transition and the need to attract private capital around government priorities, and not weaken governments' own energy transition strategies."

"No government wants to go to market saying that the scientific evidence is inconsistent with the environmental integrity of their own energy transition strategy."

The Taxonomy helps investors understand the alignment of their portfolio with the Paris Climate Agreement. It is a more sophisticated disclosure than a singular company ESG score. This is because companies are complex. An energy company may have a coal subsidiary and a renewable energy subsidiary.

The performance thresholds provide pathways for companies in understanding what's green and what's not. It supports companies in their own CapEx decisions.

In the case of an energy company, it gives us a fuller picture as to whether the company is able (and willing) to transition.

The Taxonomy also anchors conversations through the intermediation chain. An asset owner that receives pages of disclosure from one asset manager may find it hard to compare with pages of disclosure from another asset manager. The Taxonomy provides a common starting point.

For responsible investors, it's fashionable to talk down the EU Taxonomy. Some say that the Taxonomy is too ambitious, the performance thresholds too strict, the data sets too cumbersome and the results not useful.

When it was published, more than one responsible investment professional called me to ask, "but where's the Taxonomy?" It was pages and pages of activity and performance thresholds. I think many investors were expecting an off-the-shelf neat, usable tool, perhaps on a page or two.

The Taxonomy is often misinterpreted. It is a classification system to understand the extent to which an investment portfolio is aligned with the Paris Climate Agreement. And that's all. It is a disclosure tool. For the first time, there is an independently developed, regulatory tool (not an asset manager's proprietary tool) that allows me to compare the alignment of the portfolios in which I invest.

In 2020, I ran a PRI working group of over 40 investors that took steps to implement the Taxonomy to help socialise and test how it works. The case studies varied from portfolio alignment to an assessment of a single company.

The case studies demonstrated how the EU Taxonomy could be used, and also, a sense of what the results would look like.

Previously, with just a tweak or two, many asset managers were able to say their portfolios were sustainable. The EU Taxonomy has changed that. With the Taxonomy regulation now in effect, we'll see much more attention to the EU Taxonomy in the months and years ahead.

I asked Nathan Fabian whether he expected the Taxonomy to be simpler than it turned out to be.

"I think there was a view at the start that setting up the Taxonomy would be relatively easy, which was based on the assumption that lots of consumer products have energy or environmental ratings scales and surely we were just creating another one of those."

"But the Taxonomy ended up as a framework across the whole economy for everything from commercial buildings to bio-energy across six environmental objectives. And it was mandatory. When you put these parameters around the Taxonomy, it increased the complexity."

"At the time the HLEG recommendation was made, we didn't know what form the regulatory design of the Taxonomy was going to take. In Europe, there's also a huge emphasis on consultative policymaking and of course the industry needs accuracy and these factors added detail and complexity to criteria."

"There would have been ways to simplify the Taxonomy along the way. For example, we could have not required disclosure for every financial product, not required disclosure on all harms, focused on climate change only, and made the Taxonomy voluntary."

"This would have delivered a more streamlined, less detailed framework, in a voluntary space, which could have allowed time to build comfort and experience with the Taxonomy. The complexity of the framework and mandatory requirements at the start of the process put a lot of pressure on the Taxonomy."

"We were implementing a number of changes at the same time: sustainability goal aligned criteria, economic activity-based reporting, and for individual assets, companies and financial products, plus both revenue and CapEx exposure. That represents a lot of changes at the same time and that's why people were overwhelmed with the complexity."

"There are some benefits here too. We presented the fully elaborated version, and the implementation and learning challenges can be understood early, accommodated as we go (through implementation)."

In my view, the EU Taxonomy constitutes a significant step forward in disclosure and portfolio measurement, despite it's complexity.

"Of course, that also means that real economy policies will need to continue to adjust to be in line with our targets", Martin Spolc said.

"In the example of the Taxonomy, currently, we sometimes need to go beyond the existing law, the prevailing aim being alignment of the Taxonomy criteria with the ambition of the European Green Deal objectives. Also, there is a particular difference between requirements for a 'substantial contribution' and requirements for 'doing no significant harm'."

The Taxonomy helps us understand what constitutes 'substantial contribution' in a way that real economy policymaking often does not. While real economy policymaking should, in theory, align with the Taxonomy's performance thresholds, there may be complex political reasons that make alignment challenging (or even impossible).

By channelling private capital to activities that do align with the Taxonomy's performance thresholds, it will rebalance what's possible, and real economy policymaking will follow.

At least, that's what financial sector policymakers are hoping for.

SFD-Argh!

The implementation of the Sustainable Finance Disclosure Regulation (SFDR) has, I'm afraid, not worked well.

SFDR is a regulation, not a directive. A "regulation" is a binding legislative act. It must be applied in its entirety across the EU. A "directive" is a legislative act that sets out a goal that all EU countries must achieve. However, it is up to the individual countries to devise their own laws on how to reach these goals.

There was great opportunity here. Not a directive, subject to the regulatory cultures of member-states, but a European-wide regulation that applies to all European investors. It was intended to re-orient capital towards sustainable growth and to help retail and institutional investors make more informed investment decisions. It was also intended to address greenwashing.

SFDR sets out disclosure requirements for three types of funds in the regulation's Articles 6, 8 and 9. Article 6 describes the default fund. The investor must integrate financially material ESG issues, but that is the extent of their approach to sustainability.

An Article 8 fund promotes environmental or social characteristics, such as 50% lower greenhouse gas emissions. An Article 9 fund pursues a sustainability objective. For many market participants, the distinction between Article 8 and 9 funds was unclear, prompting the European Securities and Markets Association (ESMA) to write to the European Commission to say so. It's implied that an Article 9 fund prioritises its sustainability objective. In short, it's a higher bar.

Article 8 and 9 funds with an environmental objective must disclose against a benchmark, either, the Paris Aligned Benchmark, which is 50% emissions reduction and 7% reduction per year, or the Climate Transition Benchmark, which is 30% emissions reduction and 7% reduction per year.

Article 8 and 9 funds with an environmental objective must also report against the EU Taxonomy.

Article 8 and 9 funds must also disclose against a range of environmental and social indicators set out in the regulations.

For many investors, SFDR is confusing. Noting the time scales involved in European policymaking, SFDR was introduced during a period of change for responsible investment, during which investors increasingly focused on how investment can contribute to sustainability outcomes.

Most responsible investment professionals will have an opinion on how to fix SFDR.

There are positives to the regulation. It is at the entity and fund level, and so there's a degree of comprehensiveness that was previously lacking.

It requires all investments to integrate financially material ESG issues or explain why not, which, while more and more accepted as a requirement for investment decision-making, was by no means universal.

It introduces principal adverse indicators (PAIs) requiring investors to disclose the extent to which their investment activity causes adverse impact (or harm).

While PAIs are a disclosure requirement, the investor is required to consider PAIs in investment decisions. This is a step change for responsible investment because it is one of the first times that investors are regulated not on their investment processes or financial performance, but on the real-world impact of their investment decisions.

PAIs include GHG emissions, carbon footprint, investments in the fossil fuel sector, water footprint, biodiversity risks, violations of UN Global Compact, gender pay gap and exposure to controversial weapons.

And so there's much to welcome.

But unfortunately there are also many challenges. SFDR is a disclosure regulation. It is not a label but it was inevitable that it would be interpreted as a label.

European rule-making has two levels. Level 1 is the regulation. Level 2 is what's called the Regulatory Technical Standard. It sets out what's necessary to meet the regulation. Level 1 requirements started in March 2021, but the "Level 2" disclosures only in January 2023.

It is the difference between "promote social and/or environmental characteristics" and "a sustainable investment objective" which has caused considerable concern. When the Level 2 disclosures were still in development, many fund managers set out to undertake Article 9 disclosures.

But through regulatory guidance and market interpretation it became clear that, in order to qualify for Article 9 disclosures, the investment portfolio would need to be "100% sustainable", in that, every company in the portfolio would need to be sustainable.

National regulators were issuing guidance through the course of 2022, while asset managers were seeking to interpret and either change their investment strategy or change the disclosure.

There's also no definition of what constitutes "sustainable". That's left to the investors.

Given attention to greenwashing through 2022, investors became risk-averse, and began to "downgrade" their funds from Article 9 disclosures to Article 8. As such, Article 8 became such a broad category as to undermine the intention of the regulation.

Article 9 was mostly reserved for unlisted assets, or if listed, then active portfolios without a mainstream benchmark.

For active and passive portfolios with a mainstream benchmark, Article 8 could be a fund with strong stewardship, engaging companies with conviction to change and transition or Article 8 could be a fund that just excludes coal or tobacco.

It means that asset owners are back to reading prospectuses unable to compare like-for-like given the breadth of Article 8 funds.

Because asset managers are left to define a sustainable investment, each will have their own methodology, often based on a data provider's methodology, which in turn, will differ (for example, say, MSCI's SDG alignment score with, say, ISS's SDG alignment score).

ESMA has complicated this further, issuing a consultation on guidelines for the use of ESG or sustainability-related terms in funds' names (ESMA 2022).

ESMA's starting point is:

- a quantitative threshold (80%) for the use of ESG-related words;
- an additional threshold (50%) for the use of "sustainable" or any sustainability-related term only, as part of the 80% threshold.

SFDR has been costly in legal fees and goodwill. Lawyers are the winners, charging a fee for updates to prospectuses to meet the new requirements.

But that said, I don't regret SFDR. It has forced asset managers to think about their products and how they're marketed. It has forced the industry to address its conflated and confused terminology, using terms interchangeably, with little regard for how the terms are interpreted.

We're not there yet.

Not including stewardship is, in my view, Europe's big mistake, and one the UK has got right in its Sustainability Disclosure Requirements (SDRs).

Privately, some European policymakers blame investors. Investors shaped the regulation and lobbied hard for the different disclosure requirements. It's perhaps another indicator of the growth of responsible investment that in 2020, when the regulation was under development, investors lobbied for "Article 6" disclosures, but then, in 2021 and 2022, when responsible investment was well and truly established, opted for Article 8 disclosures.

Perhaps the recalibration from Article 9 to Article 8 will mean a recalibration from Article 8 to Article 6, although at this stage, it's too early to tell, and Article 8 remains too broad to be a useful distinction.

And perhaps, the industry will coalesce around terminology and disclosures that achieve the aims of the regulation even if, as the Level 2 disclosures are rolled out, it's currently very muddled.

In the meantime, there's little love for SFD-Argh.

I put this to Jon Lukomnik, asking for a US perspective of the regulation. He said, "My problem with the regulation on responsible investment right now is that responsible investment is a process, and it's become productized."

> "You can't understand the strategy by looking at portfolio holdings. But regulation is based on holdings analysis. I would prefer disclosure on intentionality and an assured, qualitative report. The first rule should be: say what you do and do what you say."

> "It is too easy to have holdings-based disclosure, and an asset manager buys Sustainalytics, ISS or MSCI ratings, and doesn't think at all and gets Article 8 status."

Fiona Reynolds had similar concerns, "I worry that there will be over-regulation that just becomes compliance that becomes box ticking, rather than bringing about change."

Philippe Zaouati said, "SFDR is a good example of demonstrating that the most important thing for a piece of regulation is the way that the market adopts it."

> "Sometimes you issue regulation, and sometimes it's useless, because the market doesn't react."

> "The big problem for regulators is that they are not reactive enough. They issue regulation and it takes years to look at it again. In a perfect world, I think regulation should be more flexible and help evolve the direction of the market."

> "The purpose of SFDR is transparency and therefore to incentivise the fund manager to report. The more you commit, the more you report. So, if you

don't do anything, you don't report, if you do ESG integration, you report a bit more, and if you do positive impact, then you report more again."

"But the fact is that these three levels have been translated by the market into a label, because we don't have any label in the market."

"Then of course, it led to big concerns at the regulatory level. If it is to be a label, then we'll need to define the different levels more precisely."

"At Mirova, we advocated to pause, saying 'let's enforce the regulation as a transparency requirement and if we need to work on regulation on labels then let's do that separately'. We shouldn't use SFDR for something that it's not intended for."

"If the European Commission were to define what's sustainable using the current Taxonomy then there would probably be only 3–4% of Article 9 funds in the market, and it would have killed the regulation itself."

"So I think it was a good decision by the European Commission not to set out a regulatory definition of what's sustainable, and rather to leave it to the market."

"But it's clearly not the end of the story. There will be a review of the regulation, we will continue to work on it, and build a usable and relevant disclosure, and if necessary label framework."

"When we were developing this piece of law," Martin Spolc told me, "investors said, 'don't be too prescriptive, give us flexibility, we know how to do this'. So that's the approach we took with our proposal. It was never meant to be a labelling regime. It was supposed to be a disclosure regime."

"But when SFDR was being implemented, many investors said, 'we don't actually know what to do, this is not sufficiently clear, we need more guidance'. And many investors started to use the SFDR for purposes other than disclosure."

"So recently, we clarified that SFDR is not a labelling regime, it is a disclosure regime."

"We are also doing a comprehensive assessment of SFDR. We have published a questionnaire for public feedback. This will prepare the ground for the next Commission to consider any changes."

"I personally think that the sustainable finance framework, including the SFDR, has all the features that allows investors to invest sustainably."

"Five years ago, we were told by investors that they would like to scale up sustainable investments but there were three obstacles in their view:

"First, the lack of clarity on what is sustainable, so we developed the Taxonomy."

"Second, the lack of data and transparency, so we developed the disclosure regime for both the financial sector and companies."

"Third, the lack of tools, so we have established the EU Green Bond Standard and the climate benchmarks. Most recently, we have put forward a legislative proposal to bring more transparency in the ESG ratings."

"We have also explained how our SF framework can be used by companies and the financial sector to support their transitioning efforts. And we have also addressed a number of usability concerns that stakeholders have identified."

"I believe that by now we have done everything that we were asked to do and we committed to."

"So it is the turn of the financial sector to take up the tools that they called for and we delivered and to bring the impact on the ground. The Commission's priority going forward will be to support both the financial sector and companies with their transition efforts."

SDR

The SDRs are the FCA's equivalent. They are similar, but not the same to as SFDR, in both content and acronym.

SDR stands for Sustainability Disclosure Requirements. In 2022, the FCA published a consultation paper setting out its approach (FCA 2022).

Line 1.1 of the consultation says: "Consumers must be able to trust sustainable investment products. Consumers reasonably expect these products to contribute to positive environmental or social outcomes."

To address greenwashing, investment products labelled as ESG or sustainable must, in the FCA's view, contribute to environmental or social outcomes. This is a step change from previous regulatory efforts that required investors to integrate financially material ESG issues.

"Some firms are making misleading sustainability-related claims about their investment products," said the FCA. "The market is difficult for consumers to navigate." "It may be challenging for a consumer to determine whether a product is integrating ESG factors into financial risk and return considerations" or "whether it has a specific sustainability goal to achieve a real-world positive impact."

With these proposals, the lens is the contribution of the investment product to real-world sustainability impact, rather than the financial performance of the investment product.

The FCA, benefitting from hindsight, intended the proposals to cover both labels and disclosures; while this is where SFDR has ended up, its intention was only disclosures.

The FCA also sought to address fund marketing rules. The products could not be marketed as ESG or sustainable, unless they conformed to the FCA's requirements. This addresses another issue with SFDR, where fund naming and marketing was addressed not by SFDR, but by subsequent consultation undertaken by ESMA.

The FCA proposed three labels: Sustainable focus, sustainable improvers and sustainable impact, with no hierarchy. In other words, all labels contribute to real-world sustainability impact, but in different ways. No label is more impactful than another.

Underpinning the labels are disclosures—for consumers, products, entities (the investment firm) and distributors. This allows the harmonisation of terminology across the intermediation chain.

To determine the label (focus, improver or impact), the FCA sets out three channels:

The first channel is "influencing asset prices and the cost of capital." The second channel is stewardship and engagement. And the third channel is "seeking a positive sustainability impact by allocating capital to underserved markets or addressing market failures."

In my view, the first and third channels are similar ("investment"); the difference is the characteristics of the companies.

The three labels of "focus", "improvers" and "impact" are explained as follows:

- Focus: "Products with an objective to maintain a high standard of sustainability in the profile of assets by investing to (i) meet a credible standard of environmental and/or social sustainability; or (ii) align with a specified environmental and/ or social sustainability theme." Here, the primary channel is "cost of capital" (or investment) although almost certainly the companies will still benefit from stewardship.
- Improvers: "Products with an objective to deliver measurable improvements in the sustainability profile of assets over time. These products are invested in assets that, while not currently environmentally or socially sustainable, are selected for their potential to become more environmentally and/or socially sustainable over time, including in response to the stewardship influence of the firm." Here the channel is "stewardship".
- Impact: "Products with an explicit objective to achieve a positive, measurable contribution to sustainable outcomes. These are invested in assets that provide solutions to environmental or social problems, often in underserved markets or to address observed market failures." Here the primary channel is also investment, but the assets contribute to solutions.

There were a few challenges with the FCA's approach, which the FCA addressed in its final proposals. For example, while conceptually, the three labels worked well, in practice, many investment products are a combination of channels. The FCA needed to allow for some flexibility so as not to curtail capital flows to sustainable investments due to overly inflexible labelling requirements.

It was unclear how to compare investment products in concentrated portfolios (say with, 40 companies or less) versus unconcentrated portfolios (for example, can an investor label a fund sustainable improver where it invests in 500 companies, but only engages, say 40 companies—even if the quality of the engagement on the 40 is high conviction).

There were some concerns over how to apply benchmarks and how to treat fund of funds. But overall, the FCA's proposals are robust and the consultative approach means that the FCA was able to refine their proposals to address shortcomings.

SDR and SFDR marked a turning point for responsible investment. We are no longer considering responsible investment to be ESG integration alone, but an approach to investment that seeks to achieve real-world sustainability impact.

This is likely to lead to a reduction in assets under management considered responsible. But a more robust approach to the investments that are responsible, with a clear focus on real-world impact, is a very welcome development. I expect we'll see other jurisdictions follow suit.

Another Look at the US

At the time of writing, there are something like 200 bills in 40 states on ESG-related topics. Some pro, some anti, all entirely political. Some relate to ESG integration, some relate to proxy voting.

The bills were initially believed to be financed by the petroleum lobby, but that appears to have ballooned to cover the gun lobby, the anti-abortion lobby, and even those opposing women on boards.

Running alongside most anti-ESG State-legislation was the SEC proposed ESG disclosure requirements for advisers and investors. The purpose, according to the SEC, was to "promote consistent, comparable and reliable information", similar in intention to the EU's SFDR and the UK's SDR (SEC 2022a). "Funds claiming to achieve a specific ESG impact would be required to describe the specific impact(s) they seek to achieve and summarize their progress on achieving those impacts."

"Funds that use proxy voting or other engagement with issuers as a significant means of implementing their ESG strategy would be required to disclose information regarding their voting of proxies on particular ESG-related voting matters and information concerning their ESG engagement meetings."

In short, three fund types:

1. An ESG-integration fund, where integration of ESG factors is no more significant than other factors.
2A. An ESG-focused fund, where ESG factors are a significant consideration in selecting investments.
2B. An ESG-focused fund, where ESG factors are a significant consideration in engagement strategies.
3. An ESG-impact fund, which pursues a specific ESG impact (Norton Rose Fulbright 2022).

The proposed rule followed another proposal that would have mandated climate change disclosures for registrants (a US term that refers to companies that file documents to the SEC). The rule proposal covered Scope 1, 2 and upstream and downstream Scope 3 "if material or if the registrant has set a GHG emissions target or goal that includes Scope 3 emissions" (SEC 2022b).

Also speaking on background, a policymaker told me that they didn't expect either rule to pass, which "may however be Gary Gensler's strategy" (Gensler was the chair of the SEC at the time). Companies were nevertheless beginning to disclose climate risks, particularly companies with overseas operations, and rule proposals, while not as effective as actual rules, may be the best we can hope for given the current politicisation of responsible investment.

US investors I spoke with had mixed views on whether Gensler was "up for the fight", adding that, the rules will almost certainly not survive the courts.

Another insider told me that the SEC understands that the Supreme Court has an appetite to curtail the authority of regulatory agencies such as the SEC, and as such, is taking great care not to give the Court an opportunity to use ESG-related rule-making as the reason to do so.

On the one hand, the Republican right's new found interest in "ESG" has had a chilling effect on responsible investment and there appears to be a chasm opening up between what US and European investors can say on responsible investment topics.

I found myself in one meeting with a US asset manager, where the head of responsible investment was disparagingly describing new approaches to responsible investment as "being designed by Europeans", while European asset managers were increasingly exasperated by their US counterparts' views on net zero.

But—when it comes to investment practice—the gap is not as wide as the rhetoric.

My experience of working with US investment managers is that the litigious culture means that, when the fund says they "do ESG", they do indeed "do ESG". There is often a deeper quality of implementation and analytical rigour than equivalent funds in Europe.

The Department of Labor ERISA rules (covered earlier, which govern US retirement plans, and include provisions on ESG integration) were always political. So much so that in March 2023 President Biden was forced to veto a House resolution that sought to overturn a DOL rule that clarified that ESG factors could be considered that are relevant to risk and return analysis (which in turn, overruled Trump-era rule-making). As a result of the political pingpong, most retirement plans have simply tracked the middle ground, and will continue to do so.

The anti-ESG movement has forced many large US asset managers (and indeed, state funds) to clarify their commitment to responsible investment, which is welcome. While some have weakened their commitment, many have strengthened their commitment.

And while the argument will continue to evolve, the Inflation Reduction Act has simply changed the economics of, in this case, environmental issues.

The Act authorises 391 billion USDs in spending, mostly through subsidies and tax reform. The Act also includes a new grant programme (or green bank) called the Greenhouse Gas Reduction Fund (GHGRF), which, among other environmental objectives, has begun to roll-out solar, particularly in lower income communities.

The US is taking a different path from that of Europe and the UK perhaps because it has the public financing opportunities to do so. Certainly mainstream European investors are more outwardly vocal and more comfortable on responsible investment topics than their US counterparts. But a large part of US institutional assets continue to be invested responsibly.

NYC comptroller, Brad Lander, who in September 2022 wrote to BlackRock setting out his expectations on climate commitments is among responsible investment's most vocal cheerleaders (NYC 2022).

Responsible investment is not Europe's alone.

References

ESMA (2022), ESMA Launches a Consultation on Guidelines for the Use of ESG or Sustainability-related Terms in Funds' Names [online]. Available from: https://www.esma.europa.eu/press-news/esma-news/esma-launches-consultation-guidelines-use-esg-or-sustainability-related-terms (Accessed, February 2023)

European Commission (2015), GREEN PAPER Building a Capital Markets Union. [online]. Available from: https://eur-lex.europa.eu/legal-content/EN/TXT/?uri=CELEX%3A52015DC0063 (Accessed, January 2023)

European Commission (2018), Action Plan: Financing Sustainable Growth. [online]. Available from: https://eur-lex.europa.eu/legal-content/EN/TXT/?uri=CELEX%3A52018DC0097 (Accessed, January 2023)

European Union (2021), EU Taxonomy Climate Delegated Act [online]. Available from: https://eur-lex.europa.eu/legal-content/EN/TXT/?uri=CELEX%3A3 2021R2139 (Accessed, February 2023)

FCA (2022), Sustainability Disclosure Requirements (SDR) and Investment Labels [online]. Available from: https://www.fca.org.uk/publication/consultation/cp22-20.pdf (Accessed, February 2023)

Le Monde (2023), Macron Calls for 'Pause' in EU Environment Regulations [online]. Available from: https://www.lemonde.fr/en/environment/article/2023/05/12/emmanuel-macron-urges-for-pause-in-eu-environment-regulations_602 6372_114.html (Accessed, June 2023)

Norton Rose Fulbright (2022), US SEC Proposes New ESG Disclosure Rules for Funds and Advisers [online]. Available from: https://www.nortonrosefulbright.com/en/knowledge/publications/915ef285/us-sec-proposes-new-esg-disclosure-rules-for-funds-and-advisers (Accessed, June 2023)

NYC (2022), BlackRock Inc.'s Commitment to Net Zero Emissions [online]. Available from: https://comptroller.nyc.gov/wp-content/uploads/2022/09/Letter-to-BlackRock-CEO-Larry-Fink.pdf (Accessed, June 2023)

SEC (2022a), SEC Proposes to Enhance Disclosures by Certain Investment Advisers and Investment Companies About ESG Investment Practices [online]. Available from: https://www.sec.gov/news/press-release/2022-92 (Accessed, June 2023)

SEC (2022b), SEC Proposes Rules to Enhance and Standardize Climate-Related Disclosures for Investors [online]. Available from: https://www.sec.gov/news/press-release/2022-46 (Accessed, June 2023)

12

Corporate Disclosure

Drivers of Corporate ESG Reporting

Corporate reporting tops any industry poll as the biggest barrier to responsible investment. More data, investors tell us, is what's needed to transition investment portfolios to be sustainable.

I asked Eelco van der Enden, CEO at the Global Reporting Initiative (GRI), to set out his views on the principles of good quality corporate disclosure.

"Good quality disclosure is comparable, verified data, prepared for a multi-stakeholder audience, which includes investors, clients, suppliers, employees and national governments."

"Without comparability, investors cannot make decisions, and you need assurance because you need to be able to trust what you read."

"When you deal with sustainability related topics, not so long ago, the data wasn't comparable and wasn't verified. But this has changed. For GRI, about 40% of companies disclosing to GRI have undergone some form of external assurance."

There are a range of drivers that explain why companies prepare ESG reporting.

1. First, ESG issues are financially relevant to the company. In almost every jurisdiction, securities law requires publicly listed companies to disclose issues that will affect the company's profitability, revenue forecasts, resilience of their supply chains and so forth.

© The Author(s), under exclusive license to Springer Nature
Switzerland AG 2023
W. Martindale, *Responsible Investment*, https://doi.org/10.1007/978-3-031-44536-1_12

2. Second, many ESG issues are regulated. There is an increasing number of regulatory reporting requirements for companies. For example, a company may be required to disclose a TCFD report, environmental footprint, such as chemical use, a modern slavery statement, human rights due diligence reporting and a gender pay gap report. For investors, distilling various corporate disclosures into decision-useful data sets requires considerable analysis, particularly for complex, multi-jurisdictional, multi-language companies, which explains investors' reliance on ESG data providers.

3. Third, most companies subscribe to voluntary reporting initiatives, including GRI and the Carbon Disclosure Project (CDP). While voluntary, it's in most companies' interests to disclose. The reporting is useful to the company, investors require it, it avoids NGO attention, and it avoids ESG ratings providers using proxies. It can also save companies time. Rather than field multiple similar, but not the same data requests, the company can refer stakeholders to their GRI or CDP report.

4. Finally, investor reporting requirements necessitate company reporting requirements. Investors are increasingly subject to ESG reporting requirements themselves. In order for investors to report, they must understand the ESG-related risks and opportunities of the companies in which they invest.

If investors do not have satisfactory disclosure there are four actions.

1. Engage the company to disclose.
2. Engage the company's regulator to introduce mandatory disclosure.
3. Divest from the company and focus elsewhere.
4. Or reprice the company (an adjusted "fair value" as sustainability risks are unknown).

For Nathan Fabian, Chief Responsible Investment Officer at PRI, we're at roughly "step 3 out of 5" for corporate ESG reporting, which is "standardised accounting through ISSB. ISSB will provide a global baseline for sustainability disclosures and that's worthwhile."

"Next, we need to link disclosures to planetary boundaries and international rights frameworks."

"The limitation with current accounting methods is most of them treat the frameworks on environmental and societal issues as open ended, and not placing any particular constraints on company business models, but that's not right. Disclosures ultimately need to link back to boundaries. Then we'd be getting a disclosure framework that will get us closer to where we need to be to inform economic transition and implications for investors."

"Is that institution there at the moment? I'd say it's only partially there."

"GRI has some of the characteristics that are required. But it's clearly going to be a combination of the different regulators that's needed to provide a trustworthy and maintainable framework that has those dynamics."

David Blood told me, "One of the issues with sustainability and ESG, is this belief that one size fits all, that we can come to a single score on ESG. Those assumptions are simply not correct. There are trade-offs in how we're thinking about businesses. You cannot come to a single score. How you weigh the different parts of that score are subjective."

"Corporate emissions reduction is built upon carbon disclosure, science-based targets and an implementation strategy that breaks the job into manageable chunks. Developing comparable data sets on impact, robust standards, and measurement and reporting norms should be the highest priority for sustainable investors. Accountability is essential."

ISSB

To great fanfare, in 2021 the various reporting groups came together. Finally.

The groups were the Global Reporting Initiative (GRI), the Carbon Disclosure Project (CDP), the Sustainability Accounting Standards Board (SASB) and the International Integrated Reporting Council (IIRC).

A new global baseline would be established by the International Sustainability Standards Board (ISSB) headed by ex-Danone chief, Emmanuel Faber.

ISSB is part of IFRS, the International Financial Reporting Standards, which has considerable clout.

ISSB recruited a number of well-established company, investor and sustainability professionals and got to work, consulting with companies, investors and stakeholders throughout 2022, establishing an overall disclosure framework and a climate disclosure framework, extending to other sustainability issues in 2023 and beyond.

Bob Eccles told me: "IFRS didn't care about this stuff [sustainability]. IFRS came out with this consultation [on sustainability] out of the blue."

"My view on this is that it's incredibly positive. You need a set of standards for ESG disclosure and it requires regulatory enforcement."

"[The IFRS foundation] knows how to develop standards that are investor relevant."

I have met with ISSB representatives across a number of forums.

On climate change, I'd characterise ISSB's proposals as "TCFD plus". By doing so ISSB has formalised TCFD as a global standard. ISSB is likely to be adopted by regulators, with early commitments from UK and EU policymakers—and other jurisdictions, such as Singapore, following suit.

Disclosures include Scope 3 emissions. Including Scope 3 reporting supports the US SEC's efforts to enhance and standardise climate disclosures. Some investors lobbied against the SEC's inclusion of Scope 3, presumably because Scope 3 disclosures would include their own investments. For the SEC's part, this helps to push back against the investors that did so.

The climate disclosure framework also includes some interesting, but cautious analysis on offsets, including where and how companies should include offsets in their reporting, as well as executive remuneration and target setting.

The framework considers impact on the environment, stakeholders and society, but where it links to financial materiality.

Corporate reporting is a mess, and the baseline is indeed just that, a baseline which can evolve.

But critics point out that ISSB's purpose is to harmonise global sustainability disclosures. IFRS is navigating the political balance of achieving a global baseline for sustainability disclosures with acceptance in jurisdictions beyond Europe, while meeting global sustainability goals.

In my view, it is too accommodating and does not yet reflect best practice. For investors, this is our best opportunity to get the global disclosures we need to achieve our sustainability goals. The ISSB's current sole focus on enterprise value is a limiting factor.

Responsible investment is about managing the systemic risks of climate change (and, as the ISSB disclosure drafts are published, other environmental and social issues). This requires disclosures not just on how sustainability issues affect the company, but how the company contributes (negatively or positively) to the sustainability issue.

This is because investors need to understand the cumulative and correlated affects of sustainability across the portfolio. The ISSB's approach is significantly lighter than the EU's, which will mean divergence is inevitable.

In August 2022, EFRAG, the European Financial Reporting Advisory Group, published its disclosure drafts (EFRAG, 2022). The drafts are comprehensive, incorporating double-materiality. The sector-agnostic drafts have since been ratified through the political processes.

Philippe Zaouati said, "I have no issue at all about what they're doing at ISSB. The concern I do have is where ISSB has put pressure on the European Commission in order to lower the level of sustainability commitment on European reporting. That's not ok. They should let the European Commission do what they think is relevant for European market participants."

"In May 2023, there were those trying to convince the European Commission to only publish delegated acts on climate change, and not on the other sustainability topics. That's not ok for me."

"I'm really in favour of European policymakers taking responsibility and moving forward, then working with ISSB and trying to find bridges afterwards, in order to make it as simple as possible for the companies to report. But we should not give up on the objectives we have on sustainability topics within the European Union."

Eccles added: "People expect too much from standards. That's where you get the whole single versus double materiality debate."

"There are no right answers. It's going to be messy. That's just the nature of this world. It's never going to be clean and done."

"But it's a really important foundation so you can do apples to apples comparison."

"GRI is the king of the hill. ISSB and GRI have a MOU. I think that's great. ISSB does single materiality stuff. To the extent to which this stuff [non-financial externalities] becomes important, the ISSB can develop a standard on this."

"The SEC is never going to come out with a general requirements standard the way you have with EFRAG. But US companies will use ISSB whatever the SEC comes out with."

"If you have ISSB nailing single materiality and collaborating with GRI, and GRI nailing double materiality [you get there]."

Eelco van der Enden, CEO at GRI, agrees. "From an operational and practical point of view, we are very close to having the sustainability data we need, with companies embracing the baseline from GRI and ISSB."

"But we have to deal with the realities of politics, which makes it more complicated. Corporate disclosure is now a political issue."

"And from politicians we need realism and guts."

"Change is hard for everyone. You will always see for any rule, any new type of legislation, that some business will object. They'll say that disclosure is too costly, that it risks disclosing business secrets, that it affects competitiveness, and that, anyway, the data that's provided is so complex that no one will understand it."

"Whether it's disclosure on carbon emissions, human rights, taxation or biodiversity, the pushback is always the same."

"On the other side, you have the ideologists, and the activist NGOs, and for them, the disclosure is never enough, it should always be more. They have this intrinsic misconception that business can only do bad, which is just not true."

"And so that's where politicians need to step in and embed legislation."

"We have a proof of concept. ISSB is your global baseline for financially material disclosures. GRI is your global baseline for impact. We each have, with IFRS, 25 years of experience. The uptake of our standards is already high. Investors already benefit greatly from it. So why not use it?"

I asked van der Enden why not embed GRI within ISSB.

"The difference in the role of ISSB and GRI is scope and stakeholder. ISSB helps us understand how enterprise value is affected by environmental, social and wider economic factors. It's in the financially material corner."

"GRI is about the impacts of business activities. It's about the effects that businesses have on environmental and social topics. In short, ISSB is outside in, GRI is inside out."

"And you need both. As everyone knows, you cannot run a profitable business in a dysfunctional society. That's what makes the combination of ISSB and GRI is so important. They are two sides of the same coin."

"I would have no problem embedding GRI within ISSB, but you have to look at the mandate of the IFRS foundation, which is about enterprise value. So you need to have a good look at the governance and organisational structures and mandates of both GRI and ISSB. But, would I be against that? No absolutely not."

And what about CSRD, ESRS and EFRAG?

"EFRAG is a local regulator. As you see with international accounting standards, at local jurisdictions, you see add-ons and changes, relevant to local contexts."

"I hope that EFRAG will stay as close as they can with ISSB and GRI as the international baseline for financial materiality and impact. It would be a pity if they went their own way. We worked with them to get as much alignment as we can."

"The easiest route will be for EFRAG to endorse ISSB and GRI as a starting point."

The deliberations will continue but it shouldn't stop responsible investment.

Philippe Zaouati said—and I agree with every word—"We have a lot of data. I'm always a little bit upset when I listen to all these people say we cannot invest responsibly because we lack data. That's something very strange, because we have a lot of data."

"Of course, it would always be better to have more comparable data, more standardised data, more stable data, and more established methodologies."

"But our philosophy at Mirova is that we use the data we have, that we can do a lot with the data as it is, and where we lack data, we roll up our sleeves, and try to build it. This is what we did with carbon foot-printing 7 or 8 years ago, this is what we did for data sets on biodiversity and, in 2023, this is what we did to establish data sets for Scope 4 data."

"We say to the market what we need and we incentivise the data providers to provide it."

"More and more, we will get corporate sustainability data from other sources, not just the companies themselves, but from third-parties. Technology and artificial intelligence will play a very important role here."

With corporate reporting there is also a practical limits question. The unstated thesis seems to be that if we have enough reporting investors can measure it all, monitor it all and manage it all. This clearly isn't possible.

Reference

EFRAG (2022), Public Consultation on the First Set of Draft ESRS. [online]. Available from: https://www.efrag.org/lab3 (Accessed, February 2023).

13

Impact

Impact Investments

Roger Urwin, Global Head of Investment Content at Willis Towers Watson, explained to me "For a very long time, responsible investment was all about integrating ESG into risk and return but it is now transitioning to something with impact in it. There's now a paradigm shift in the investment world to impact based on a system of wider stakeholder values."

"Is responsible investment defined that way a legitimate concept? Yes, but it's how it's executed that is the controversial part within interpretations of fiduciary duties."

A new PRI and UNEP FI programme, a Legal Framework for Impact, tries to answer this question. But while impact is understood as relevant to all investments, it is worth first understanding impact investment as an asset class.

Impact investments are often, although not exclusively, in unlisted markets, and impact fund management, is typically higher fee than other forms of responsible investment.

This is because impact investments can be subject to complex investment structures, in hard-to-reach geographies, with associated reputational risks, and challenging measurement requirements. Investors will want regular reporting of their impact. And often impact will be intangible and secondary (for example, the benefits to the families of those in employment).

But impact investments also have many advantages. From a financial perspective, they are a source of uncorrelated returns. In other words, in

W. Martindale, *Responsible Investment*, https://doi.org/10.1007/978-3-031-44536-1_13

periods of market stress (and we've had several tail events in recent years), many impact investments will be unaffected.

In the case of social housing, impact investments may outperform during a period where other asset classes under-perform, with high demand and therefore high occupancy rates.

And impact investments provide investors with clear, tangible, real-world impact. Often in an impact investment, the investor is more directly linked to the real-world impact of the investment than other forms of responsible investment.

In my experience, impact investment is often the "exciting" bit of responsible investment. The impact is clearer, the travel is more exotic, the storytelling is more real, and the projects, particularly new technologies, are more interesting, even, exciting.

One impact manager I looked into undertook research into the future of travel, accurately predicting the uptake in electric scooters and electric bikes, identifying the cities where scooters and bikes are more likely to become a common mode of transport (supportive politics, younger but affluent workers, not too geographically dispersed, perhaps not too hilly, stressed public transport), identifying the most promising start-ups, and then investing. This, to me, is a fascinating approach to investment.

Defining impact is not straightforward. GIIN and IMP are the two go to frameworks.

For GIIN, it's "impact investments are investments made with the intention to generate positive, measurable social and environmental impact alongside a financial return" (GIIN 2019).

In 2020, a book by Sir Ronald Cohen caught my eye. "Impact: Reshaping Capitalism to Drive Real Change" (Cohen 2020). The book includes a number of interesting case studies that span Cohen's career. But it is also a demonstration of the challenges involved in defining what we mean by impact investment.

"The biggest of all the asset management firms, BlackRock, which has nearly $7 trillion under management, is confident that impact investing is the future", Cohen claims. Here, Cohen's wrong. This is simply not BlackRock's dominant approach to investment decision-making.

Some institutional investors, however, have invested in impact investments.

A good example is Strathclyde's direct impact portfolio. Strathclyde is Scotland's largest local government pension fund, with assets under management of 27.6 billion GBP and just under 270,000 members from across the West of

Scotland, administered by Glasgow City Council (Stathclyde Pension Fund 2022).

The direct impact portfolio was introduced following the global financial crisis in 2009 to provide financing to small and medium-sized businesses impacted by the withdrawal of bank financing.

The portfolio grew—first set at £300 million, then 3% of net asset value, and in 2018 a target of 5%, which was around £1.4 billion. The remit also changed to include impact investments with the vast majority of investments in the UK, including affordable housing.

For many years, the UK has had an under-supply of affordable housing, in particular, housing for social rents. Councils tend to require developers to build a minimum quota of affordable housing in order to secure planning application.

Social housing makes for an attractive investment. Tenants tend to be long-term, some rents will be paid for through housing benefit (and therefore, relatively secure), and, as mentioned above, demand for social housing tends to rise in periods of economic stress.

And so the question becomes, why impact investing is such a small part of the investment landscape. The reasons include high fee, often riskier asset classes and less liquidity, but also, there's a perception issue. In my experience, many pension fund decision-makers are still of the view that if you're getting something (impact) then you're giving up something (financial returns).

And because impact investments are often in unlisted markets, the risk profile may not work for all investors.

I asked ShareAction's CEO, Catherine Howarth, whether we should legislate for minimum allocations to impact investment.

Howarth said, "I'm intuitively cautious about legislating for impact. Indeed, I'm uneasy about political interference in pension investment decisions full stop. We're seeing that in the US with the anti-ESG crowd dictating how state pensions are invested. It's deeply concerning."

"Rather than dictate from above, I'd prefer that members of schemes be consulted on, amongst other things, their appetite for impact-driven investments. You may well get a strong yes in many schemes. Then you have a mandate from the people whose money this is. Put the members at the centre of the system. That's a better way forward than assuming you, as regulator or politician, know what's best for them."

"I think trustee discretion is a good thing."

Impact investment is an important incubator for responsible investment. Among other topics, responsible investors can learn from impact investors in impact reporting.

But, while growing, it's likely to remain a relatively small part of responsible investment.

PRI and Freshfields on Sustainability Outcomes

The terms impact and outcomes are increasingly associated with responsible investment, not as an asset class distinct to responsible investment, but core to responsible investment itself.

The PRI's terminology on outcomes is as follows (PRI 2020):

"All investor actions shape positive and negative outcomes in the world."

"Issues such as human rights abuses, climate change and inequitable social structures seriously threaten the long-term performance of economies, investors' portfolios and the world in which beneficiaries live."

"Expectations from beneficiaries, clients, governments and regulators over how investors should respond have changed—driven by increased visibility and urgency around many of the SDGs."

"To support meeting the SDGs, investors must understand how they can increase the positive outcomes and decrease the negative outcomes arising from their actions."

PRI does differentiate between "impact" and "outcomes" but the difference in my view is too subtle to warrant explanation, and so I use the terms interchangeably.

The shift to impact and outcomes is significant. With the exception of financing of illegal activities, such as money laundering or tax evasion, the financial sector is largely regulated on process, not outcomes.

Investors are asking themselves whether an explicit focus on real-world sustainability impact is consistent with fiduciary duties.

In 2020, PRI and UNEP FI with support from The Generation Foundation commissioned law firm Freshfields to answer this question.

The Fiduciary Duty in the 21st Century programme concluded in October 2019.

At the time, the PRI had no mandate for impact, but to obtain a mandate for impact, it needed to start working on impact.

In response, the report "A Legal Framework for Impact" proposes an elegant distinction in investment for sustainability impact between what Freshfields calls "instrumental" and "ultimate ends" (Freshfields 2021).

"Instrumental investing for sustainability impact is where achieving the relevant sustainability impact goal is 'instrumental' in realising the investor's financial return goals."

Although the investor seeks real-world impact, the motivation to do so is financial.

"Ultimate ends investing for sustainability impact is where achieving the relevant sustainability impact goal, and the associated overarching sustainability outcome, is a distinct goal, pursued alongside the investor's financial return goals, but not wholly as a means to achieving them."

Here, the investor pursues impact in its own right. Others would consider "ultimate ends" to be concessionary investing. In other words, financial returns are concessionary to real-world sustainability impact (even if, the returns are competitive, the primary objective is impact).

The report found that in all jurisdictions "instrumental" investing for sustainability impact is permitted and arguably required. The legal perspective on "ultimate ends" is less clear.

It falls to advocates of responsible investment whether legal change to enable ultimate ends investing for sustainability impact is worthwhile. It comes back to what objectives we think are considered "proper" for investors and whether those objectives need to be concessionary in ways that disrupt the constraints of existing legal models. My own view is that as soon as you introduce concessionary investing (in institutional capital), the question becomes 'how much concession?', which is not a practical decision-making framework for an investment decision-maker; nor do I think it is necessary. If the purpose is more sustainable capital markets, then investment decision-makers can pursue sustainability outcomes through the lens of instrumental investing for sustainability impact, something many investors are already doing.

Roger Urwin said "… there is a 'saying doing' gap. And, perhaps even more so, the 'saying doing feeling' gap. People feel that they want to do something. They say something. But they do something rather different. That's where implementation is weak."

"You see a lot of organisational intent [with respect to impact]. But you don't see a whole heap of implementation."

"We're in a far more challenging place where people's values now matter to investment."

"We've grown up in a world where risk and return where everything. Where values and ethics were background points."

"But in today's society, the investment industry has to be more respectful of its wider impacts, both intended and unintended."

"We've got a new form of 'impact' coming, where we think about risk, return and impact. That's the shift in the next decade and financial regulation needs to support it. Every investment portfolio is going to have to account for its impact."

In the UK, the Financial Markets Law Committee was yet again asked to look into fiduciary duties. In its 2023 paper, "Mobilising Green Investment: 2023 Green Finance Strategy", the government explained (HM Government 2023):

"We acknowledge decisions around investing and systemic risks are complicated and that trustees would like further information and clarity on their fiduciary duty in the context of the transition to net zero."

"To address this, we are taking the following steps:

a. DWP will examine the extent to which their Guidance is being followed in late 2023.
b. This will be complemented by a working group of the Financial Markets and Law Committee (FMLC) where participants, including DWP, will consider the issues around fiduciary duty and what further action is needed.
c. We will be holding a series of round tables later this year to engage with interested stakeholders on how we can continue to clarify fiduciary duty."

I expect similar studies will follow and my expectation is that we'll learn little more than what's already set out in the Legal Framework for Impact report.

References

Cohen (2020), Impact: Reshaping capitalism to drive real change, Morgan James Publishing.

Freshfields (2021), A Legal Framework for Impact [online]. Available from: https://www.freshfields.com/en-gb/our-thinking/campaigns/a-legal-framework-for-impact/ (Accessed, November 2023).

GIIN (2019), Core Characteristics of Impact Investing. [online]. Available from: https://thegiin.org/characteristics/ (Accessed, January 2023).

HM Government (2023), Mobilising Green Investment: 2023 Green Finance Strategy. [online]. Available from: https://assets.publishing.service.gov.uk/government/uploads/system/uploads/attachment_data/file/1149690/mobilising-green-investment-2023-green-finance-strategy.pdf (Accessed, June 2023).

PRI (2020), Investing with SDG Outcomes: A Five-Part Framework. [online]. Available from: https://www.unpri.org/download?ac=10795 (Accessed, May 2023).

Strathclyde Pension Fund (2022), Investments. [online]. Available from: https://www.spfo.org.uk/index.aspx?articleid=14446 (Accessed, May 2023).

14

Another Look at Stewardship

Stewardship 2.0

Stewardship, which comprises engagement and voting, is the way in which investors engage the assets they own, their stakeholder groups, regulators and policymakers, and how they vote at company AGMs.

Stewardship has evolved considerably in recent years, with increased focus on collaborative stewardship.

By owning a share of a company I get a say in how it's run. Every year, companies ask my view (and that of all their owners). I can vote to elect the chair, the board of directors and vote for or against management on resolutions, including on remuneration packages.

I can, for example, use my vote to tell companies that they need to do more to tackle climate change. But as one voice of many, typically, I'm unsuccessful, which explains why many investors collaborate.

The most well-known collaborative initiative is Climate Action 100+, which includes many hundreds of investors engaging the world's largest emitters.

In 2019, PRI described stewardship as failing, "yet remains our best hope". It's a view I agree with. In this chapter I will take another look at stewardship, starting with CA100.

© The Author(s), under exclusive license to Springer Nature Switzerland AG 2023
W. Martindale, *Responsible Investment*, https://doi.org/10.1007/978-3-031-44536-1_14

Climate Action 100+

Climate Action 100+ is a collaborative engagement initiative organised by PRI, IIGCC, IGCC, AIGCC and Ceres.

Although the investors that collaborate vary in their conviction on climate change-related topics, the three core asks have wide support, and CA100 has had a number of successes.

The three asks of companies are to:

1. Implement climate change governance with board accountability.
2. Set a net zero target, near-term targets and action to meet the targets.
3. Prepare and publish a TCFD report.

The organisations behind CA100 identified the 100 (at time of writing, increased to 166) companies with the highest GHG emissions. More than 700 investors have since signed up to participate, including a mix of asset owners and asset managers. The total AUM of the investors is around 70 trillion USDs.

In the early stages of the initiative, each company was assigned a lead investor (or sometimes, lead investors) and a number of supporting investors.

The allocations were determined based on the investors' portfolio holdings and the geographies in which they're based and have engagement staff.

With limited stewardship resources, the divide and conquer approach made sense, allowing an investor to get to know and build deep relationships with one or two high polluting companies, knowing that other investors were doing likewise with other high polluting companies.

But in a move that frustrated NGOs, who does what was unpublished. This was because stewardship was typically undertaken in private. The argument put forward was that investors needed to build trusted relationships with companies and companies needed confidence that the meetings would stay confidential.

The problem, however, was that other investors, clients, regulators and stakeholders found that without the public disclosure it was hard to determine substance over form. Attempts were made by NGOs and journalists to find the information and publish anyway.

The lead investor's role was (and where the engagements are ongoing, is) to coordinate meetings between the company and the supporting investors, propose an agenda and run the meeting.

The meetings I participated in tended to be cordial. One of stewardship's challenges is that in our professional (and, I assume, personal) lives we tend to avoid conflict. But the company representatives may well have a different position to that of their investors, and so conflict cannot be avoided.

And the meetings I participated in tended not to be decision-making. The company representatives may include investor relations staff, sustainability staff and, occasionally, management or even board members.

The representatives will present the unique challenges faced by their industry, set out the progress they've made (often, compared to selected peers), acknowledge that they want to go further ("if only policy makers would do more to help"), and then set out their plans for the months and years ahead.

If a company does implement a more advanced climate change strategy, the company is unlikely to want to attribute it to shareholder engagement.

Some investors put forward senior stewardship professionals that have done their homework, ask tough questions and are focused on outcomes. This can lead to high quality exchanges, and the company representatives will relay pressure to their superiors, which may bring forward action.

However, other investors put forward less senior or less capable stewardship professionals that have not done their homework, are there to box tick (for example, the number of CA100 meetings attended in a year), and generally contribute little more than some notes for their internal reporting systems.

While the investor groups provide reporting materials, the quality of the individual engagements depends very much on the lead investor.

However, despite these issues, CA100 provides us with a good practice case study of collaborative engagement. It is surprising that CA100 has not been replicated across a broader range of sustainability issues. In late 2022 and early 2023, that began to change with new collaborative initiatives on human rights and nature underway.

But for all its success, investors and stakeholder groups are starting to think about what's next for CA100.

One spin off is the Dutch Climate Coalition, which goes further than the asks of CA100, calling for oil and gas companies to "prove your commitment to Paris", boost low carbon solutions, explain how natural gas acts as a transition fuel and do not use the high oil prices to increase oil investments (MN 2022).

And in a November 2022 article on Responsible Investor, ShareAction's Peter Uhlenbruch argued in favour of minimum standards for CA100 participating investors (Responsible Investor 2022). At the moment, you can sign up to CA100, but there's little scrutiny on the quality of your contribution. For many asset owners that are not direct owners of the CA100 companies (instead, investing via a third-party manager) there is limited opportunity to participate.

Other potential developments included increased transparency, such as investor members' voting decisions, expanding asks to include political activity of target companies, capital expenditure targets, and a deny debt engage equity strategy, whereby investors align their fixed income allocations with their CA100 engagement.

In 2023, both BP and Shell weakened their decarbonisation commitments. I'm sure BP and Shell did consult with some of their shareholders, but as I understand it, not via CA100.

I asked Nathan Fabian, Chief Responsible Investment Officer at PRI, whether this was a threat to responsible investment.

"The recent revision of BP's decarbonisation targets do represent a threat to responsible investment as a concept, but it's the same threat that responsible investors were already aware of. That policy making doesn't always align with planetary boundaries and international frameworks on rights."

"I think this is understood within responsible investment. We've known about it for over 20 years on climate policy. Perhaps in recent months, with energy supply issues arising from the war, the need to keep the heating on through winter, and pandemic recovery it's been a more acute set of issues than we've been used to dealing with previously. But it's already an inbuilt assumption into responsible investment that policy will fluctuate."

In June 2023, CA100 announced Phase 2. On the surface, Phase 2's updates appeared modest. New thematic and regional engagements were introduced. Perhaps, most materially, expectations for lead investors and individual engagers to disclose voting records on CA100-related votes.

I asked Stephanie Pfeifer about CA100's future.

"When talking about Climate Action 100+ I think it's important to firstly recognise its impact to date and how much progress has been made. Without question the initiative changed the conversation in relation to the world's largest corporate greenhouse gas emitters and the role and importance of investors in corporate engagement".

"It has also delivered results. When Climate Action 100+ launched at the end of 2017, only five focus companies had set net zero emissions by 2050 or sooner commitments. As of October 2022, three quarters of focus companies have now set a net zero emissions by 2050 or sooner ambition that covers, at least, their Scope 1 and 2 GHG emissions."

"However, we recognise focus companies collectively still need to go further and faster to support global efforts to limit the temperature rise to 1.5 degrees Centigrade, starting with credible transition plans. Moreover, the case for further and faster corporate action on climate change has never been more compelling."

"Looking ahead, building on the foundations laid during the first five years and the lessons learned, I'm optimistic that Climate Action 100+ will continue to have a positive impact between now and 2030. In particular, a number of key developments have been introduced for the second phase of the initiative, including a shift in focus from corporate climate-related disclosure and commitments to the implementation of corporate climate transition plans."

Engine No 1

The best known recent example of a high conviction collaborative engagement was led by Engine No 1 (who do not call themselves "activists").

Californian-based asset manager, Engine No 1, was established by hedge fund manager Christopher James with $250 million of his own money.

Small, but well-resourced, in December 2020, Engine No 1 shot to fame having penned a letter to Exxon, commenting on its financial performance, decades of climate change denial, and proposing four independent directors join the company's board. In just over a decade, Exxon had lost around two-thirds of its value.

What made this remarkable was that Engine No 1 held just $40 million of Exxon's company shares, or around 0.02%.

The letter "proposed a new path forward", with four asks: "Refresh the board. Impose greater long-term capital allocation discipline. Implement a strategic plan for sustainable value in a changing world. Realign management incentives" (Reenergize Exxon 2020).

The third ask included "more significant investment in net zero emissions energy sources and clean energy infrastructure" and "ensure the company can not only set Scope 1, 2, and 3 carbon emission reduction targets, but also make them part of a sustainable, transparent, and profitable long-term plan focused on accelerating rather than deferring the energy transition."

Exxon's directors and management had repeatedly ignored requests from shareholders to publish a decarbonisation strategy, unlike Exxon's European peers. Given its role in the US economy, its size, and its GHG emissions, Exxon became a predictable target for shareholder engagement.

Following the publication of the letter, Engine No 1 began a campaign—and campaign is the right word—to change Exxon's directors. The hedge fund started to speak to other shareholders, including the US's largest pension funds, CalPERS and CalSTRS.

Anders Ruenvad, one of Engine No 1's proposed directors had deep understanding of "how renewable energy companies with growing markets and declining cost curves are transforming the energy industry". Another, Kaisa Hietala is "a trained geophysicist and environmental scientist".

CalSTRS almost immediately announced its support for the resolution (CalSTRS 2020). In a press release, CalSTRS said: "CalSTRS intends to support Engine No 1's alternate director slate for ExxonMobil to equip the board with the relevant skills and experience to strengthen long-term performance, resiliency and strategic positioning."

Others followed suit, citing Exxon's unwillingness to engage on climate change and Engine No 1's focus on financial materiality. In an interview, CalSTRS's Aeisha Mastagni said, "it's hard to think of a better [example of 'activist stewardship'] given the company's financial performance and decades of indifference to its shareholders" (HBR 2021).

With CalSTRS on board, pressure turned to the big three: BlackRock, Vanguard and State Street, reported to own 18% of Exxon's shares between them. To be successful, Engine No 1's proposed directors would need 50% of the vote. The arguments were sound, the momentum was clear, and, in the end, the big three indicated their support for Engine No 1's directors.

Engine No 1 calculated the probability of success with one of the big three, then two, then all three supporting the proposals, and going into the vote, was confident of two directors. In the end, Engine No 1 won three.

In a HBR article on the campaign, Bob Eccles and Colin Mayer concluded "it could be the prelude to many similar [campaigns] by other funds to the benefit of shareholders and society at large" (HBR 2021). Certainly I thought that too.

But Engine No 1 has not repeated its campaign. I would imagine it doesn't need to. When Engine No 1 comes knocking, a company would be minded to listen.

But neither have other asset managers followed suit.

Given many asset managers' stewardship teams are now regularly collaborating, through initiatives such as Climate Action 100+, and the many half-baked decarbonisation strategies published across the oil and gas sector, why there was no "shareholder spring" is a question worth asking. As is the question of why it was an unknown, small, start-up investor, and not a much bigger asset manager with deeper pockets and a team of stewardship experts. I think there are a few answers.

When I met with Engine No 1 in London, the team was quick to present their motivations as financial. In a podcast, "Capital Isn't", James argued that his motivation was financial materiality, not ideology (Capital Isn't 2021). In the US, too many sustainability issues get "bogged down" in ideology, he said.

Engine No 1 was in the right place (in the US, this couldn't have been achieved by an overseas investor), right time (Exxon's poor financial performance and unwillingness to accept the reality of climate) and right message (focus on financial materiality).

CalSTRS's early support provided credibility.

Engine No 1 was not a competitor to the big three and as such, perhaps easier for the big three to get behind.

Stewardship was in the spotlight. Asset managers were investing in their stewardship capabilities. Asset owners were beginning to more fully assess stewardship when deciding on a new mandate.

Engine No 1 was prepared to finance the campaign. Engine No 1 spent $12.5 million, but had budgeted $30 million for the campaign.

Nevertheless, $12.5 million is still a lot of money for an asset management industry under considerable fee pressure, which may explain why others haven't followed suit. It was certainly a gamble.

Engine No 1 also had the luxury of focus. For an asset manager with 100s, or in some cases, 1000s of shareholder meetings to assess, sole focus is just not possible. Some resolutions are too prescriptive. Engine No 1's letter was brief and factual.

There was clear CEO-support. James decided to "do battle" with Exxon following a family dinner (WSJ 2021).

Perhaps another factor was the unique political dynamics. The Trump administration's withdrawal of the US from the Paris Climate Agreement led to high expectations on US companies and US investors to drive change, in the absence of action by politicians.

In an interview with me, Bob Eccles added:

"I think it's really hard to do it [what Engine No 1 did]. This was not an environmental campaign. The newspapers keep writing it up that way. It was an economic campaign."

"There's not that many activist investors out there. Most don't have this bent for it."

"But you have to ask yourself the question, how many really high profile companies are there where there's a very strong economic thesis behind the campaign? It's hard to find them."

"You've got to have the skills of an activist, know how to run a campaign, put together a slate, it's not a common skill set. There are maybe 50 activist hedge funds, maybe 10 that really matter."

The campaign catapulted Engine No 1 from just another start-up asset manager to almost celebrity-like status, but some asset managers, perhaps most, are not chasing the limelight on an issue that is still perceived as divisive.

Split Voting

As well as engagement, there's also increased focus on voting, and more recently, split voting. In 2021, the UK DWP commissioned a non-binding study on split voting, which appears to have had DWP support (DWP 2021).

Asset owners that pursue low cost market exposure tend to do so through pooled funds, administered by an asset manager.

Asset managers can offer passive pooled funds for as little as 1 or 2 basis points in fee (and often lower). In theory, if you can't sell (because it's passive), you should engage (if a company's in the index, it's in the portfolio, regardless of whether the investor is concerned about a company's ESG-related risks). The only way to mitigate those risks is through stewardship.

Asset managers provide stewardship of varying quality. Stewardship is assessed by an asset owner, or the asset owner's consultant, in the selection of the asset manager.

Stewardship is typically undertaken centrally. It's not feasible, nor in my view desirable, for an asset manager to provide engagement on one topic for one asset owner client, and on another topic for another client or to present two opposing views to a company where asset owners are not in agreement.

The challenge of navigating multiple client views in engagement becomes even more challenging for large asset managers when deciding how to vote, because, whereas engagement is often in private, voting is public. An asset manager could be investing on behalf of both a red-state US client and a European client, with opposing views on a US energy company's decarbonisation strategy. The asset manager must vote in the same way for both clients, each with their own set of beneficiaries and stakeholders.

One response is to introduce split voting. The asset manager will have its default view, but provide clients with the option to override that vote with their own view if they disagree with the default.

BlackRock is one of the asset managers to have introduced split voting. A June 2022 press release said that nearly half of BlackRock's global index equity assets are eligible for split voting. BlackRock says that it is "in response to growing client interest" (BlackRock 2022). My take is that BlackRock was between a rock and a hard place, and could not meet the expectations of its diverse client base, and so split voting is a convenient solution (and many other asset managers have followed suit).

Although some asset managers have provided split voting for some time, split voting has gained in appeal very recently including for pooled funds (where multiple asset owners invest in the same fund). BlackRock offers a range of options. Clients can select from a menu of third-party proxy voting policies, clients can simply take control of all their votes (although, this would have a cost implication), or clients can continue with BlackRock's default voting choices.

Accompanying the increased interest in split voting is technology that allows the end saver to signal their preferences. The best known company (at least in the UK) is Tumelo. Tumelo's website says, "Tumelo enables shareholders to have a voice on the issues they care about by providing Expression of Wish and pass-through voting technology to fund managers, brokers, and institutional investors" (Tumelo 2023).

The arguments in favour of pass-through voting are as follows:

It shortens the intermediation chain, votes can fully reflect asset owners' (and perhaps savers') views, companies may be more likely to respond to asset owners (as the ultimate owner of their company) and most asset owners' liabilities are longer-term than their asset manager mandate. In the case of Tumelo, with app-based technology for savers, it integrates often disenfranchised savers in the voting process.

But against, asset managers may shy away from the difficult decisions.

Take an asset manager with 10 asset owner clients, two of which are highly engaged on sustainability topics, 8 of which are somewhat agnostic, or perhaps lack the resources to form a view. The two engaged clients are likely to incentivise higher conviction voting on behalf of all 10 clients.

With split voting, asset managers may take a step back, allowing their more engaged clients to vote as they wish. Either asset managers believe the issue matters, or they do not, and understanding that is an important part of the manager selection process.

It risks asset manager investment in voting systems, not stewardship. And the UK pension market is characterised by comparatively small pension funds without the resources to consider hundreds and potentially thousands of votes (that's why they pay their asset managers a fee). Asset owners could of course contract their own voting provider, but that comes at a cost.

In the case of Tumelo, savers that repeatedly put forward a view, only for that view to be ignored may be further disenfranchised. Tumelo might argue that this would incentivise the saver to invest their pension with another provider instead. Tumelo is now working directly with institutional investors to achieve vote consistency.

I put this to academic and investor, Jon Lukomnik, an adviser to Tumelo. He said, "You have to distinguish between pass-through voting and expression of wish voting. Pass-through voting also impacts smaller pension funds, endowments and foundations which aren't large enough to use separately managed accounts."

"I don't think it splits ownership and control from economic risk. I actually think it increases the legitimacy of voting. There are attacks in the US on the voting power of large index funds—part of it is political, part of it is just anti large entities, part of it is Chamber of Commerce—the corporate viewpoint saying the stewardship teams of BlackRock, State Street and Vanguard have too much control."

"But, with pass-through voting, what is actually done is empowering and legitimising the votes from the large index funds. It's savers saying, 'yes, this is the conscious choice, we want you to vote this way'."

"If I'm in charge of stewardship at one of the big three, one way to do it would be to assume that the pass-through voting is a reflection of my total universe, not my most passionate voice. In fact, perhaps it's safest to mirror vote."

"In other words, if I were the head of stewardship for an asset manager with some amount of pass-through or expression of wish voting, I would use the statistics from it as an important data point. So, not dispositive, but influential as to how I voted. And I do think it increases accountability."

While I find Lukomnik's arguments compelling, I'd also favour:

1. Pre-vote disclosure (or principle-based pre-vote disclosure, but preferably priority votes, upfront) in which asset owner clients are consulted.
2. Published escalation measures (escalation measures may include votes against management on a particular issue, votes against a director or directors, or filing a resolution).
3. Post-vote outcome-based disclosures (the real-world change achieved by the vote) and that, in the most difficult of voting decisions, asset managers make the call.

Sometimes asset managers may make the wrong call, but I think on balance, I'd rather that, than they don't make a call at all. I think there's logic in vote consistency as a short-term intervention, but asset managers that repeatedly vote contrary to my beliefs should prompt me to stop investing with that manager, not override their voting. On split-voting, I'm split. I could probably be convinced either way. But—in the long-term at least—if I had to vote, I'd vote against.

Sovereign Engagement

Another topic that needs to be handled with care is sovereign engagement.
   I would consider there to be three categories of sovereign investor:

1. Investing in sovereigns to hedge interest or inflation risks. Here, an investor may have a series of choices between a physical government bond, a derivative government investment, such as a swap, or a supranational agency bond, or local authority bond that may have similar interest rate and inflation characteristics as the national government.
2. Investing to seek interest rate and inflation exposure, and potentially currency exposure, through overseas sovereigns, typically sovereigns with strong credit ratings, such as the US, Europe, France, Germany, Japan, UK, Australia or Canada.
3. Investing to seek investment returns through a portfolio of government bonds, including emerging markets, with flexibility in selection.

Despite the size of the sovereign bond market, ESG assessments are likely to be a relatively small contributor to investment decision-making. Geopolitics, security and issues like national finances are likely a far more significant consideration than issues.
   This may change, as climate change-related weather events become more prevalent and countries fail to adapt, or social inequality leads to civil disruption, but even still, in the first category, which by AUM is the largest of the three categories, the investor has little choice but to invest in their home country.
   And the sad truth about climate change is that it is likely to disproportionately affect poorer countries, often countries that have made little contribution to climate change.
   Often investors are more influential as economic actors within their home market than as owners of overseas debt.

There are also uncomfortable ethical issues here. Investors using their influence to engage governments subject to democratic processes is ethically complex. Indeed, it may be counterproductive.

I could imagine that the Bolsonaro, Modi or Trump governments would have reacted negatively to overseas investors engaging on environmental or social topics in their countries.

And if the overseas government ignores the investor, then would the investor withdraw their capital, which (theoretically) would raise the costs of borrowing to be paid for by the local population? It's not a trivial decision.

A project attempting to answer these questions is ASCOR, "Assessing Sovereign Climate-related Opportunities and Risks", headed by the UK's BT Pension Scheme and Church of England Pensions Board.

The bit I find particularly interesting is how to address perverse incentives (pulling capital from the countries that require it).

The project has put forward a publicly available assessment tool to support investors in engaging sovereigns and understanding the transition risks and opportunities.

Stewardship Resourcing

A defined contribution pension fund may charge its customers around 50 basis points, or half a percent, in annual fees. In other words, if I invest £100 in my pension fund, I pay 50p in fees. Hypothetically, let's say that of that 50 basis points, there are 20 basis points of administrative costs, things like a call centre, annual benefit statements, office costs, the website and marketing.

That leaves 30p for every £100 I invest that is spent on the investment strategy.

A pension fund might hedge investment risks, such as interest rates and inflation, investing in low-risk, defensive assets. Say this costs 5p of my 30p, and, is applied to, say, £40 of my investment.

That leaves 25p to spend on my remaining £60, and here, the investor would pursue a growth strategy, expecting a return of a few percent depending on, in the case of a pension fund, age of the saver, risk appetite and geography—some markets tend to have higher return targets than others.

A specialised fund manager charges as much as 100 basis points in fees. Some charge more.

To continue with our example, this means I could put, say £15 of my remaining £60 investment in a fund, where the fee is 100 basis points. This would cost 15p of my remaining 25p in fees, leaving me 10p in fees for the rest of my £45. And so, to invest within my fee cap, perhaps I'll select a low cost passive investment product with stewardship.

Although, the asset manager may say that you're getting stewardship, in practice, we can see from the level of fee, it's unlikely the pension fund is contributing much to the asset manager's stewardship programme.

This is the challenge facing pension fund trustees. Where can you make savings? On stewardship. Do I invest in index fund 1, that costs 25 basis points, and includes high conviction, well-resourced stewardship? Or do I invest in index fund 2, that costs just a few basis points, and doesn't really do stewardship? The composition of the portfolio and the portfolio's returns would be the same.

The issue becomes more complex when you start to think about who should pay for stewardship, or more accurately, which type of saver. Stewardship suffers from the tragedy of the commons.

Let's say pension fund A is managing money on behalf of supermarket workers. On the whole, supermarket workers are not well-paid.

Why should pension fund A use their savers' hard-earned, hard-saved pension fund investments to pay for expensive stewardship when it's not just pension fund A that will benefit from it, but all pension funds?

In the race to the bottom on costs, stewardship has become an optional extra. In my experience, few pension fund trustees will select an asset manager based on the quality of their stewardship. Pension funds will require minimum levels of stewardship. But stewardship is not yet the differentiator it should be.

Let's continue our example. Of the 10 basis points, let's say—and I think we're being generous here, that 5 basis points are spent on stewardship (of my £100 investment, 5p is spent on stewardship), the other 5 basis points are spent on the fund manager's administration costs.

Of that 5p, my experience is that the stewardship includes, in descending order (in other words, most of the expenditure is highest in the list):

- Asset managers' portfolio managers (the individuals that make investment decisions) engagement of companies, mostly on "for information" or governance themes
- Proxy voting, outsourced, plus oversight
- Company-specific stewardship (because the company has unmanaged sustainability risks)
- Thematic stewardship (for example, on biodiversity)
- Collaborative initiative systemic issues stewardship
- Public policy engagement

My experience is that stewardship is a fraction of a pension fund's fee basis and the highest impact stewardship activities (e.g. systemic stewardship and policy engagement) are the least funded within that.

Modern portfolio theory tells us to diversify, and therefore most of our returns are the average of market performance. But to enhance market performance, the two most effective levers are the ones that receive a fraction of my fee.

Regulators have sought to address this. They've introduced principle-based codes and disclosure requirements—and I expect they will continue to do so.

For all those that argue engage don't divest (as former pensions minister, Guy Opperman did on a PRI-hosted webinar in March 2022, "if you believe in divestment, get off this webinar now"), then engagements need to be properly resourced (PRI 2022). And right now they're not.

Claudia Chapman told me, "Everyone is always lamenting how little resource for stewardship there is."

"My experience of those working in stewardship and responsible investment—both on the policy and regulatory side, as well as the asset owner and manager side are committed, well-informed, have wide knowledge and understanding."

"They work hard and are stretched thinly. I think this is improving—though may wax and wane a little—with the current economic and US political environment. Is it any different in any other profession?"

"The [UK Stewardship] Code has influenced the increase in resources. But if investors are as committed to their stewardship objectives as they say they are, and we believe stewardship to be an effective tool to achieve change, then the profession needs to be better resourced. More people with the right training, skills, knowledge and experience."

"In the meantime, we need to establish, through both academic evidence and the sharing of experiences, what is working, what is not working. Be realistic about the limitations of stewardship and focus efforts where they can have the most impact."

In February 2023, PRI and the Willis Towers Watson Thinking Ahead Institute (TAI) established a working group to "better understand current stewardship practices, resourcing requirements and other key cost" (TAI 2023).

The group found that stewardship resourcing is a fraction of investment costs despite its importance.

Roger Urwin told me "Stewardship is not well-resourced and not well delivered."

"There is a business model paradox stalking the industry. We invest massively in portfolio management resources picking stocks and asset classes—where much of the effort is the negative sum in alpha."

"And we don't invest enough in stewardship where the value is a positive sum, particularly the systemic stewardship at the industry and public policy level."

"We are not sizing or shaping our stewardship effort at all well and we will need to increase our resources and focus for the sustainability challenge ahead, particularly on climate and net zero."

I asked Urwin if the government should set minimum criteria for stewardship. "I'd love to have governments be more prescriptive but I don't see it happening. I believe in the 'art of the possible'. I don't see governments doing things that are substantially prescriptive."

"On a practical basis, net zero has profound significance to countries and non-state actors. Legislation should set out what net zero means for non-state actors. Regulators should step in to make it easier for them to play a role. You need to take away all the obvious frictions."

"Telling pension funds what to do when they have fiduciary duty to do something in a different way, makes prescriptive stewardship difficult."

"If you do it well, and remove the barriers, and put a safe harbour in, you're building a better system."

Escalation

One of the biggest challenges for stewardship teams is knowing when and how to use escalation. Companies may be slow to respond to engagement, not respond at all, or simply disagree with whatever it is that the investor is seeking to achieve.

Escalation is sometimes considered binary: If your engagement is unsuccessful, you can divest from the company. But there are a series of steps investors can undertake without divestment, steps I believe that, if done well, are more powerful than divestment.

This includes:

- Votes against a chair or board member
- Votes against a CEO's remuneration packages
- Proposing a new chair or board member who has expertise relevant to the investor's engagement objective

- Votes in favour or against resolutions relevant to the investor's engagement activity
- Filing or co-filing shareholder resolutions. While resolutions may be advisory, and receive minority votes, they are an important way to raise the profile of an issue that is otherwise unaddressed
- Engaging specifically on dividends, and whether dividends should be recycled into CapEx to address sustainability considerations
- Selling (or not buying) the companies debt, while retaining the company's equity
- Engaging NGOs, media and other stakeholders in support of an issue
- Engaging policymakers. Requiring disclosure on the issue may prompt the company to take action
- Supporting legal action, this is particularly the case, where a board member or company director is considered to have breached their legal responsibilities (by, for example, failing to implement an energy transition strategy).

Most investors make use of proxy advisors to support their voting decisions. The proxy advisor's default recommendation is based on what they perceive as the best interests of the shareholder, which, by definition, can take a narrow view.

It may be in the best interests of an oil company to delay decarbonisation activities, but it is unlikely to be in the best interests of the portfolio or end investor.

Investors are beginning to consider voting across their portfolio, not just each company in isolation.

Roger Urwin told me "Proxy voting is getting much more significant and interesting, but it's still an obscure part of the playing field. It's not easy to understand all the motivations of the different participants in proxy votes and how to interpret them. Each participant has its own unique take. But this will grow as stewardship and engagement develops."

One escalation measure that's worth paying attention to is legal action. In February 2023, UK law firm, ClientEarth filed a lawsuit against Shell's board of directors (ClientEarth 2023).

This was the first time directors were being sued in a personal capacity, ClientEarth saying that their failure to meet climate targets put the company at risk.

A range of investors supported the claim, including UK pension fund, NEST.

In a press release, Mark Fawcett, NEST's chief investment officer, said, "Investors want to see action in line with the risk climate change presents and will challenge those who aren't doing enough to transition their business" (NEST 2023).

Lawyer friends of mine dismissed the case as a publicity stunt, but I expect, in some form, legal action will be a feature of high conviction stewardship.

Deny Debt

One example of an escalation strategy is to deny debt.

Lothian Pension Fund is one of Scotland's largest pension funds with £8 billion in assets under management. In June 2020, Lothian Pension Fund adopted a new policy, "engage your equity, deny your debt".

In Lothian's 2020 stewardship code submission, the pension fund says, "As an organisation, we've made a commitment that we'll not provide any new financing to companies which aren't aligned with the goals of the Paris Agreement on Climate Change. While the trading of equities (shares) doesn't affect the capital position of a company, subscribing to new bonds and new equity does provide companies with funding."

The primary market is where stocks and bonds are created (in the case of stocks, via an initial public offering), the secondary market is where stocks and bonds (and other financial instruments) are traded. If I sell, someone buys. The company is unlikely to care too much.

For a company that has already sold its equity, divesting does not directly affect the company's cash flows. Potentially, if lots of investors sell, and far fewer buy, then share price will fall, affecting the company's cost of capital. But there are few sustainability issues that prompt enough investors to sell.

Richard Roberts, Inquiry Lead at Volans, said to me, "There should be a much bigger focus on primary markets. The way we sometimes talk about secondary markets materially impacting a company's access to capital, is a misleading characterisation."

"The only time that companies really raise money from stock markets is when they IPO, after that secondary market investment is largely a form of speculation."

"I think we should have much more focus on primary markets, much more focus on bond markets, rather than just equity markets—the balance has, to date, been the other way round."

"In secondary markets, voice is more interesting than exit. I support the ideals, values, aspirations and ambitions of the divestment movement, but as a tool for having impact, I don't think divestment is a particularly effective one."

Fossil fuel companies are typically considered income stocks, a company that pays regular, and often high, dividends. Income stocks differ from growth stocks, which may not pay a regular dividend, but are expected to grow, such as a newly public technology firm.

For as long as we use fossil fuels in the real economy, fossil fuel companies are likely to provide their investors an income.

Fossil fuel companies that have little prospect of transitioning (perhaps due to their location) should pay higher dividends, returning capital to investors, rather than reinvesting it in new oil and gas exploration. Fossil fuel companies also tend to perform well in periods of high inflation.

This is why some investors find it hard to exclude fossil fuel companies citing fiduciary duties as a barrier. The argument's circular. Not enough investors sell, and so company share price is unmoved, and so not enough investors sell.

But the fossil fuel companies may also ignore investor stewardship.

Lothian, and Lothian are not alone, think they have an answer. Instead of selling equity, sell debt. Or more precisely, deny debt. Don't buy it. The difference between equity and bonds is that bonds have a fixed term, or maturity.

In May 2015, Shell issued a fixed rate 10-year bond paying interest of 3.25%. It is likely that, on maturity (in May 2025), Shell will re-issue a 10-year bond paying whatever at the time is the going rate of interest. The bonds are issued on the primary market.

Unlike its share price, Shell will care about the level of interest because it directly affects cash flows. Higher interest rates mean higher costs, which means lower profits.

The advantage of denying debt, rather than equity, is that the investor retains its stewardship rights through its equity investments. The investor can continue to engage unsustainable companies, file or co-file resolutions and vote at shareholder meetings.

When deploying this strategy, investors should be selective. Lower rated companies will be more affected than higher rated companies because there is more investor demand for more credit-worthy companies.

My experience is that equity portfolios and credit portfolios tend to be run in isolation, even within the same asset manager.

A credit portfolio manager will be responsible for a credit fund, an equity portfolio manager for an equity fund. While, the investment strategy may be similar, and make use of the same data sets, it is difficult, and potentially, not legal, for an investor to pursue a change objective across both its equity and credit funds.

This is because it's likely that the asset owners investing in the credit funds will be a different set of asset owners to those investing in the equity funds. And the asset manager will have a responsibility to all its clients.

Asset managers could decide to exclude more companies from its credit portfolios (or have a higher bar for exclusion from its equity portfolios). But the strategies will often be marketed independently.

These problems are not insurmountable particularly for asset owners that invest themselves, rather than outsource to asset managers. But for those that do, it's likely to require more thought to get the approach working well.

A strategy could be, "we like this company, we think the company has potential, but its management is not adequately addressing sustainability issues. As such, we'll invest in the company's equity, and use our voting rights to seek to the change the company's management, while denying the company debt, using our fixed income portfolios to push up the cost of capital, and put financial pressure on the company to transition."

Assuming the strategy is successful, I would be comfortable buying the company's debt, and benefiting from any resulting financial out-performance through my equity investment.

References

Blackrock (2022), BlackRock Expands Voting Choice to Additional Clients. [online]. Available from: https://www.blackrock.com/corporate/newsroom/press-releases/article/corporate-one/press-releases/2022-blackrock-voting-choice (Accessed, February 2023).

CalSTRS (2020), Statement on Alternate Board Members for ExxonMobil. [online]. Available from: https://www.calstrs.com/statement-on-alternate-board-members-for-exxonmobil (Accessed, February 2023).

Capital Isn't (2021), The Engine No. 1 David vs Exxon Goliath With Chris James. [online]. Available from: https://capitalisnt.com/episodes/the-engine-no-1-david-vs-exxon-goliath-with-chris-james-CoxW_uCL/transcript (Accessed, February 2023).

ClientEarth (2023), ClientEarth Files Climate Risk Lawsuit Against Shell's Board with Support from Institutional Investors. [online]. Available from: https://www.clientearth.org/latest/press-office/press/clientearth-files-climate-risk-lawsuit-against-shell-s-board-with-support-from-institutional-investors/ (Accessed, February 2023).

DWP (2021), The Report of the Taskforce on Pension Scheme Voting Implementation. [online]. Available from: https://assets.publishing.service.gov.uk/govern

ment/uploads/system/uploads/attachment_data/file/1018751/taskforce-on-pen sion-scheme-voting-implementation.pdf (Accessed, February 2023).

HBR (2021), Can a Tiny Hedge Fund Push ExxonMobil Towards Sustainability?. [online]. Available from: https://hbr.org/2021/01/can-a-tiny-hedge-fund-push-exxonmobile-towards-sustainability (Accessed, February 2023).

MN (2022), Investor Statement Dutch Climate Coalition. [online]. Available from: https://www.mn.nl/blog/183/investor-statement-dutch-climate-coa lition (Accessed, May 2023).

NEST (2023), Nest's letter to Shell [online]. Available from: https://www.nestpe nsions.org.uk/schemeweb/nest/nestcorporation/news-press-and-policy/press-rel eases/Nests-letter-to-Shell.html (Accessed, November 2023).

PRI (2022), A Legal Framework for Impact: Investing for sustainability impact in the UK [online]. Available from: https://www.unpri.org/policy/a-legal-fra mework-for-impact-investing-for-sustainability-impact-in-the-uk/9639.article (Accessed, November 2023).

Reenergize Exxon (2020), Letter to the Board of Directors [online]. Available from: https://reenergizexom.com/materials/letter-to-the-board-of-directors (Accessed, February 2023).

Responsible Investor (2022), Big Read: Investors Set Out Wishlist for Second Phase of CA100+. [online]. Available from: https://www.responsible-investor.com/big-read-investors-set-out-wishlist-for-second-phase-of-ca100/ (Accessed, February 2023).

TAI (2023), Thinking Ahead Institute and PRI to Create New Global Standard for Stewardship Resourcing. [online]. Available from: https://www.thinkingaheadin stitute.org/news/article/thinking-ahead-institute-and-pri-to-create-new-global-sta ndard-for-stewardship-resourcing/ (Accessed, February 2023).

Tumelo (2023), Voting Technology? It's All We Do. [online]. Available from: https:// www.tumelo.com/ (Accessed, February 2023).

WSJ (2021), The Hedge-Fund Manager Who Did Battle with Exxon—and Won. [online]. Available from: https://www.wsj.com/articles/the-hedge-fund-manager-who-did-battle-with-exxonand-won-11623470420 (Accessed, February 2023).

15

Systems Thinking

Where Next?

Many responsible investment professionals see responsible investment's future in impact investments, new metrics, new disclosures and new sustainability themes, as well as understanding responsible investment's limitations, and being clear on responsible investment's objectives. I would subscribe to this too. These are all important developments. And for many investors, this is rightly their focus; it is impactful, it matters, and their work is making a difference.

But for me, thinking about the role of responsible investment more broadly, I'd like to see further focus within the industry on how we think about systems change.

Systems change gets substantially less attention in the responsible investment literature, and this is why, this book concludes by taking a look at systems change, and what investors (or perhaps, more broadly, the responsible investment industry) can do about it.

"You have to understand the present and future in the context of the past. Investment and indeed economics grew up with physics envy. So it got cute by solving the mathematics of markets ahead of human challenges" Roger Urwin told me.

"But we're evolving to a new paradigm where real-world impact matters. Behaviours in capital markets were narrowly defined by shareholder primacy. That's changing and that's a fundamental shift."

"We've started to define the ground-rules of what we need to do to be multi-stakeholder oriented."

© The Author(s), under exclusive license to Springer Nature Switzerland AG 2023
W. Martindale, *Responsible Investment*, https://doi.org/10.1007/978-3-031-44536-1_15

I asked Urwin how investors might think about systems change.

"The model of the system I use is captured in the 'STEEPLE' mnemonic: social technological economic environmental political legal and ethical. That's the ecosystem. It wasn't like that in the past. It was about economics. So systems thinking at its first level is that everything in this list connects. We've got to respect that."

"In practical terms, thinking more widely about all these elements and their connections has the potential to be turned into new methods and strategies and in turn into new solutions."

"Systems thinking leads to systems leadership. Systems thinking with strong leadership seeks to solve the toughest problems on shared terms. It sets out solutions that are not short-term fixes. It's cognisant of the complexities, the second and third order effects, the ripple effects, and the need for coalition building so that you solve the problem at the system level but you get the benefit of that at the individual level because you're part of the system."

Systems thinking is defined as a holistic way to consider all the interactions that could contribute to an outcome, thinking through how one set of actions could influence another, in order to get to a desired outcome.

If the outcome is a more sustainable financial system we need systems change and to understand how to get there we need systems thinking.

Forum for the Future describes systems change as "where relationships between different aspects of the system have changed towards new outcomes and goals. And it's driven by transformational, not incremental change" (Forum for the Future 2019).

The systems work I was involved with was PRI's Sustainable Financial System (SFS) programme launched in 2017. The work wasn't well socialised nor was it well understood. For many years PRI had prepared integration guides: How to integrate ESG issues in fixed income or how to integrate ESG issues in private equity.

To its members, the SFS programme was not contractual. There was no guide for investors, no case studies, no opportunities to showcase good practice. The SFS reports were a heavy read. Some investors considered the work outside PRI's remit. Many more just weren't aware. "If PRI works on something like this, well that's for them, but it wasn't relevant to PRI's signatories' increasingly busy day jobs" was the feedback.

I asked Nathan Fabian, Chief Responsible Investment Officer at PRI, whether he perceived the SFS programme as a success. "The programme was at the earlier end of this trend of explicitly focusing on systems thinking and was by a non-traditional actor (PRI)."

"But I think the work undertaken by UNEP Inquiry, PRI, many leading signatories in the UK, Netherlands, Nordics and France, the EU and HLEG—the HLEG is a classic financial systems mandate, even including topics such as capital requirements—were all influential on progressing towards systems thinking."

The SFS programme identified nine conditions that need to be addressed to transform the financial system (PRI 2017b).

The conditions were not necessarily negative or positive, but observations or characteristics. The conditions identified were (and I would argue, still are):

1. Short-term investment objectives
2. Attention to beneficiary interests
3. Policy maker influence on markets
4. Capture of government policy by vested interests
5. Influence of brokers, rating agencies, advisors and consultants on investment decisions
6. Principal–agent relationships in the investment chain
7. Cultures of financialisation and rent-seeking in market actors
8. Investment incentives misaligned with sustainable economic development
9. Investor process, practices, capacities and competencies.

To achieve a sustainable financial system, we need to address short-term investment objectives, but to address short-term investment objectives, we also need to address principal–agent relationships, investment incentives and investment processes. Short-termism is both cause and consequence of several other conditions.

Condition seven is interesting: "Financialisation results in the financial system paying less attention to considerations such as the value of a clean environment, decent work or economic health."

Identifying rent-seeking as a condition was quite the leap for an organisation that is mostly funded by asset managers and service providers. But here too, PRI is right. The link between saver and user of capital is ever more complex, ever more financialised and ever more subject to incremental fees through the intermediation chain.

In isolation, each part of the savings process has social worth, but in aggregate, it should not cost what it does to match provider of capital with user of capital.

Conditions 8 (misaligned incentives) and 1 (short-termism) combine. If we take a pension fund with 30-year liabilities, it's likely the pension fund invests with an asset manager across a three-year mandate. The asset manager will report its results every three months. The asset manager's management team will likely assess performance as often as every week, or even day.

The asset manager will say that they "think" long-term, but I think that's hard, when subject to short-term performance management.

The SFS programme was probably ahead of its time. It never got the traction it deserved. And PRI was probably the wrong institution to lead such a programme. PRI was beholden to its members. It is a service organisation, providing services to its members for a fee. The SFS programme sought to define the very purpose of the financial system.

But if responsible investment is to deliver real-world outcomes, a systems approach is an important part of understanding where next.

Following the publication of the SFS programme, the PRI put resources into understanding the role of asset owners, investment consultants, and what it calls, driving meaningful data. But the work is not yet at scale.

There are other groups looking at system change. Well-funded US groups like Rocky Mountain Institute undertake research and analysis on a series of topics related to the energy system.

But generally, I'm underwhelmed and I think perhaps the natural home is a policy institution. A group like the International Platform on Sustainable Finance is a strong candidate. The Platform has clear links to policymakers and is independent of private sector financing, but it would need to be better resourced and free from political oversight.

When it comes to the future of systems reform, Volans Inquiry Lead, Richard Roberts, is cautious. "There's probably been a good decade of work on financial systems reform, most of which hasn't led to anything really tangible."

"There are two approaches to systems reform: Lots of quite concrete areas that investors and other financial institutions can get involved in, looking at specific pieces of regulation and how can they be reformed in order to enable flows of capital."

"It isn't systems change, but it is engaging with regulators. Those kind of conversations are starting to, in a relatively short time frame, move the needle."

"But that's different to reforming Bretton Woods. Intellectually I love these conversations. But we haven't yet seen any tangible outcomes."

Nathan Fabian said, "The ideas put forward by the SFS programme have got traction. We're seeing lots of system actors take some level of systemic approach, whether it's central banks, whether it's the EU or even the G20—they're all showing explicit and implicit programmes on systemic approaches to climate change in particular."

Roger Urwin agrees, "The development of new ideas in the industry and implementation feels glacial. But if you look back at the past five years, we've made progress. It's an industry that has enough understanding to go forward with systems thinking. Right now, it's just not institutionalised, it's not evenly dispersed."

"In workshops I've run on horizon scanning, one of the key questions has been, in the next 10 years, do you think of the application of systems theory is critical to success. The answer has become an overwhelmingly yes."

For Alex Edmans, it is governments, not investors, that should take responsibility for systemic risks. "We do have people who do that. That's the government. It's the government's responsibility to look across everything."

"No matter how much of a universal owner you are, you will only look at the companies within your portfolio."

"Nobody owns everything. Are you going to be caring about people in, say, the Global South?"

"There are indeed systemic effects but I think that for many of these systemic effects governments are the best place to look at this. Why? Because many of these effects are more than just company- wide effects."

"Where I do think universal ownership can be beneficial is where it leads to best practices across the industry."

"A good model is the Norwegian Sovereign Wealth Fund."

"They have published various position papers on a range of ESG issues."

"They're improving general practices across the industry. And why can they do that? Because they invest in lots and lots of companies so they get really informed as to what are the key issues, microplastics for example."

But for me, a more expansive interpretation of responsible investment that includes systems change should be part of responsible investment's future, and I think there are five areas of work that seem to be gaining traction. They are:

1. Acting as a universal owner and understanding the limitations of modern portfolio theory
2. Engaging in systemic stewardship
3. Democratising the savings industry
4. Maximising use of influence
5. Political engagement.

I'll explain each in turn.

Stay Sharpe

Universal ownership theory states that large institutional investors with diversified portfolios reflect the economy. Their interests align with the public at large (Universal Owner 2020).

In 2011, a PRI and UNEP FI report, based on research conducted by Trucost, concluded that: "Large institutional investors are, in effect, 'Universal Owners', as they often have highly diversified and long-term portfolios that are representative of global capital markets."

Also in 2011, Roger Urwin, writing in the Rotman International Pensions Management Journal, said "The core idea of a universal owner is a large institution investing long-term in widely diversified holdings across multiple industries and asset classes, and adapting its investment strategy to these circumstances" (Urwin 2011).

But Thinking Ahead Institute follow-up analysis, published in 2020, found that "In practice, most large asset owners currently find factors not to manage their funds in line with universal ownership principles by either not seeing themselves as large enough; not having the long-term orientation; or not having the leadership buy-in to operate this way" (TAI 2020).

Acting as a universal owner is to address systemic issues, such as climate change or inequality, where the universal owner pursues change objectives to limit the (often negative) financial implications of the issue. This is because universal owners cannot invest away the costs of unpriced externalities.

While this is most applicable to larger investors, smaller investors can collaborate with their peers to benefit from scale.

A PRI paper on the SDGs said, "For a universal owner, environmental costs are unavoidable as they come back into the portfolio as insurance premiums, taxes, inflated input prices and the physical cost associated with disasters."

"Social concerns, such as poverty and inequality, can lead to societal and political unrest and instability, which can also create costs that will reduce future cash flows and dividends" (PRI 2017a).

Part of the reason universal ownership is not yet mainstream is explained by the limitations of MPT.

For 70 years, Modern Portfolio Theory (MPT) has defined investment.

The core of MPT is diversification. On average, introducing risk to an investment portfolio will increase returns. I could put all my money in

Bitcoin or since its IPO, Manchester United Football Club. If I'd timed it well, I'd have made money, more money than if I'd invested in the stock market at that time. Equally, if I'd not timed it well, I'd have lost money.

Nobel Prize-winning economist Harry Markowitz (and common sense) tells us to avoid idiosyncratic risk, or as the saying goes, avoid "having all of your eggs in the same basket."

But that wasn't always the case. When Markowitz published his research in The Journal of Finance in 1952, 92% of the investment market were retail investments (Markowitz 1952).

In other words, 92% of investments in the capital markets were made by individuals—mum and dad investors, rather than institutional investors. "Many small investors doing their own thing", as author and investor, Jon Lukomnik puts it. "Diversified portfolios were not the norm" (Lukomnik 2022).

Markowitz found that portfolio diversification can lower risk per expected unit of return. If I have £1000 in savings, I could buy £1000 of one company's share (say, MANU) or I could invest in a fund that buys £1 each of 1000 companies' shares.

Pre-packaged diversified investment products such as mutual funds, available to the everyday saver quickly followed.

I don't own any single company's shares, but my parents do. A company or two they "liked the look of" or perhaps one that was sold off by the government.

Instead, I invest in mutual funds. At low fees, I benefit from broad market exposure, and limit the downside risk of any one company losing value. Today, 90% of investments are made by institutional investors (pension funds, insurance funds and mutual funds).

MPT measures risk through volatility, typically through standard deviation, a measure of the degree of variation in a distribution. In the case of an investment, it is how often the price diverges from its average.

MPT is underpinned by another theory, the Efficient Market Hypothesis (EMH), which states that when there is new information, that new information leads to new pricing, that the new information is immediately reflected in the price of the asset.

And MPT is underpinned by another metric, the Sharpe ratio, named after economist William Sharpe, who shares Markowitz's Nobel Prize. The Sharpe ratio is a measure of return per unit of risk, where the unit of risk is the investment's volatility compared to a risk-free asset (an investment assumed to have no risk). The higher the ratio, the better the return per unit of risk.

MPT, EMH and Sharpe ratios are mathematical tools used to govern much of the capital markets today.

I spoke to Jon Lukomnik as research for this book. Lukomnik is co-author of "Moving Beyond Modern Portfolio Theory: Investing That Matters."

Lukomnik said, "We wrote [the book] because we saw market driven investment caring about climate change, caring about diversity. We wanted to understand why, from a financial point of view."

Lukomnik was addressing responsible investment from the other direction. Not, can investors consider issues such as climate change? Rather, why is it that investors are considering issues such as climate change?

But although investors are considering climate change, in aggregate the markets are not considering climate change at anything like the extent they should be, asset owners are not acting as universal owners.

Lukomnik continued, "People had taken modern portfolio theory and made it into modern investing theory. MPT wasn't able to explain what investors were doing with regard to systemic risks."

"One of the things that MPT and associated theories such as EMH do is they divorce MPT from the real world and the feedback loops back to investment. By doing that you get elegant math. That's all MPT is. Just math. Math about a set of market opportunities at a moment in time. That it does well."

"But by doing that, you've taken a tool that looks at a specific moment in time, to extract an efficient portfolio, and if you apply that to your investment thesis you get a compositional fallacy, because there are feedback loops."

"There's a whole body of academic literature that people look at to inform security selection, but the vast majority of variability in return is due to non-diversifiable systematic risk, such as the price level of the market."

"If you've invested recently, you understand that. From 2010–2020 it was pretty hard to lose money. In 2022 it was pretty hard to make money."

"MPT says you can diversify the idiosyncratic risk of an individual security. But it gives you no methodology to diversify or mitigate the system risks, which depending on the academic study you look at account for 75—94% of your returns."

"In other words, you're basically saying, I will only focus on the risk return profile of 6–25% of my portfolio. But if you can't diversify that 75–94% what can you do to improve the overall risk and return of the market?"

"You can try to mitigate risk in the real world. Risk and return are priced in the capital markets, but made in the real world. So we find that there's a reason investors care about issues like climate change."

"Institutional investors who own 500, 1,000 or 10,000 securities under-stand their returns will be determined by the general state of the economy, much more than their ability to pick securities."

"Part of the problem is we've all been trained to think about everything on a relative basis. But the purpose of investment is to offset your liabilities, to buy a house or pay for your retirement. Those aren't relative returns."

The models do not adequately incorporate a company's future prospects in a resource constrained world, one in which sustainability issues are increasingly relevant to a company's business model.

The models measure aggregate asset-price volatility and seek efficient fron-tiers, where return is maximised per unit of risk. The models do not care for externalities.

However, most universal owners are diversified.

But due to another theory, the tragedy of the commons, investors are not incentivised to undertake activities that benefit the market as a whole, despite that being the predominant contributor to the investor's returns.

Here Lukomnik is an optimist.

"There is a free-rider issue. Or to turn it on its head, we see a tragedy of the commons problem. I think we can minimise that for a few reasons."

"First, stewardship is relatively cheap to do. How much does it cost to join an industry initiative or to join a sign-on campaign. It's almost like paying an insurance premium. If I devote 10 basis points of my fees to stewardship and I believe that good stewardship can increase the overall price level of the market, and I have an asset-based fee, then I come out ahead."

"Second, increasingly there is a view from asset managers that when index-ation rises and fees decreases, you can't distinguish yourself by your product, so you distinguish by your stewardship."

"Third, there are also those that are socially responsible and impact investors that are motivated by trying to have impact."

While stewardship as practiced may be relatively cheap, if stewardship is to address systemic issues, my view is that current practice is inadequate. I put this to Lukomnik.

"You are increasingly seeing some adaptation of system levels thinking. For example, Aviva's approach to macro stewardship or Franklin Templeton hiring some of the best stewardship talent in the world. That said, it's still early days for stewardship."

"There are too many individuals working in stewardship that couldn't tell you about capital structure or balance sheets. The skill-set needs to combine corporate specific analysis with system analysis."

"We're in evolution. Rarely do things move monotonically. There's a reason we have the phrase two steps forward, one step back. But the trend line is in the correct direction."

Despite Lukomnik turning MPT on its head, MPT is starting to gain attention as a hindrance to responsible investment.

A final challenge for addressing MPT's limitations is that sustainability groups are not incentivised to look into MPT. Unpacking MPT is not nearly as glossy, as say, financing research on biodiversity.

"Most people with jobs in responsible investment are working for asset managers or asset owners. The focus of these institutions is their individual portfolios. Wider systems risks can easily get ignored, especially when investment firms fixate on alpha, which many do", Catherine Howarth told me.

"There are too few people that have a job that allows them to work on systems issues. It's not what most people working in responsible investment are paid to do."

For more on addressing MPT, Jon Lukomnik and James P. Hawley are the go to source. Their book, Moving Beyond Modern Portfolio Theory: Investing That Matters addresses MPT's shortcomings and charts a new course in MPT's history, that of "beta activism" (Hawley and Lukomnik 2021).

I share (what I interpret to be) Lukomnik's view, that we will not redefine MPT by seeking to redefine MPT. Rather, we address some of the contributions to MPT's prevailing interpretation, and the behaviours it drives.

Recalibrating asset owners to act as universal owners is part of that.

Systemic Stewardship

According to Bob Eccles, systemic or macro stewardship "expands on a traditional view of stewardship as the safe-guarding and nurturing of assets. It adds to this concept investors' intentional commitments to preserve and enhance the fundamental social and environmental systems that underpin the wealth-creating potential of these assets" (Forbes 2022).

"As a universal owner there's a logic that you'd do macro stewardship" Richard Roberts told me. "I would like to see lots of investors adopt the macro stewardship idea. For that to work, there has to be a recognition that if you want investors to put resource into macro stewardship, there has to be a viable business case for doing that."

I'd define systemic stewardship as pursuit of an outcome through use of influence on a systemic issue. A systemic issue is one that pervades economy and society, such as climate change or inequality.

For David Blood, "We think of systemic stewardship as anything that looks after the system upon which the long-term generation of returns depends, rather than just focusing on things that impact the near-term returns of one's own portfolio."

"So, at Generation we:

- set goals for portfolios that require system change—like net zero or an end to deforestation
- work collaboratively to create changes across the economy through initiatives like NZAM, GFANZ and Climate Action 100+
- engage on public policy in support of a sustainable economy
- fund NGOs in support of a sustainable economy through our foundation."

And for Claudia Chapman, systemic stewardship is "big-ticket stewardship". "Increasingly, it is influencing policymakers, regulators, governments and others, to change systems and markets so they encourage and support sustainable investment. This should be about directing capital towards investments that accelerate economy-wide transition."

One such example is The Investment Integration Report (TIIP), which sets out a six-step process to address systemic issues (TIIP 2022).

They are:

1. Investors must set a system-level goal that affects their investments across all asset classes.
2. Investors must decide where to focus and commit to addressing a certain systemic issue, like income inequality.
3. After deciding where to focus, investors must allocate their assets accordingly.
4. Investors can extend conventional investment tools to exercise system-level influence.
5. Investors can further leverage advanced techniques designed specifically to manage systemic risks and rewards.
6. Investors should evaluate their results towards their stated goals.

For some investors, systemic stewardship is already underway.

Aviva Investors uses the term "macro stewardship", which Aviva defines as follows:

Macro stewardship is "the idea market participants have a responsibility to help preserve the integrity of the whole financial system, keeping it in healthy service of society and the planet."

"This should be done by engaging with regulators, policymakers and many other change-makers. It is complementary to the more familiar practice of micro stewardship, which focuses on engagement with companies and issuers" (Aviva Investors 2022).

This is similar to (but perhaps, a more deliberate articulation of) the approach articulated by the UK FRC that speaks to sustainable benefits for the economy, the environment and society.

It is a more deliberate "inside out" strategy, that focuses on a change objective, and then deploys leverage (through individual company engagement, regulatory engagement and partnerships with NGOs and stakeholder groups) to make progress towards real-world sustainability impact (set out as a goal).

For Steve Waygood, Chief Responsible Investment Officer at Aviva Investors, "There are certain sustainability issues, such as antimicrobial resistance, over-abstraction of water, exploitation of natural resources, that will harm economies, leading to major economic consequences."

"Focusing on climate change, at some point there will be too much risk in the system, potentially unprecedented levels of migration, implications for security, at a certain point market integrity starts to collapse."

"Within understandings of fiduciary duty, it is recognised that investors have a duty to maintain the integrity of the market. Not because it's the right thing to do, but because if the market can't be trusted, then the market will collapse."

"This extends to sustainability issues that could harm the ability of the market to function."

"The business model of the insurance business starts to collapse, which has existential implications for the capital markets."

"This isn't just a financial stability Minsky moment—i.e. a rapid repricing of securities—this is institutional infrastructure and business models ceasing to work because premiums can't be afforded or insurance companies won't underwrite. It is a collapse in the structural integrity of markets."

"Very few people understand this. It's not just a tragedy of the horizons, it's a tragedy of perceptions too."

"The critiques of woke capitalism, are supposed to be pro growth and pro market. They think it is an immaterial add on and a luxury, where in reality it is deeply material and fundamental to long-term growth, price stability and market integrity."

I think we want stewardship to evolve as follows: Properly resourced collaborative, systemic stewardship on sustainability themes, with clear real-world sustainability goals, working across the intermediation chain, with clear escalation measures, engaging companies, regulators and stakeholders in pursuit of the goals.

Nathan Fabian said, "Systemic stewardship is definitely part of responsible investment's future, and of particular interest to PRI, as we look to address the role our signatories can play in system change. We're thinking through whether and how signatories could set targets on the type of activities they should be undertaking so we can get precise about how signatories should respond to systems change, and how we calibrate the role of the PRI."

Nick Robins added, "PRI's future is perhaps more in system engagement, looking at what are the rules of the game in the investment world that are militating against long-term returns but also the health of the system."

"In the 2000s, PRI was going with the grain of globalisation and the liberalisation of capital. But clearly that's no longer the case. We're seeing a fragmentation of markets and global capital flows are shrinking."

"At a time when the Global South needs a massive boost in terms of private investment for climate and wider sustainable development, the PRI is in pole position to examine how institutional capital could flow into emerging markets and developing economies on a routine basis."

I asked Steve Waygood why more investors aren't engaging on systemic stewardship.

"There's a freerider problem. Why should they pay for it when other people might? It's a non-excludable public good."

"You could say that about microstewardship too. That's why the regulatory responses (stewardship codes) were created. That has to be the next stage for systemic stewardship."

"So you have a freerider problem, there's a lack of regulatory oversight, there's a lack of recognition that there is a duty there."

"In listed equity, you're the ultimate regulator of the business, it's not the same in sovereign debt."

"There are ethical issues too. Is it legitimate for financial institutions to seek to engage with policymakers and politicians? Up until now, it's been financial institutions talking their own book."

"The lack of engagement at system level is itself a market failure."

And how do we address it?

"It's about standards, it's codes, it's regulatory clarity, it's market demand, it's investment consultants [including in their selection criteria]."

"And the reason why? ESG integration is not enough, therefore, micro stewardship is not enough either."

"If you support shareholder resolutions that are value eroding you're exposing yourself to legal liability. If macroeconomic issues harm GDP, harm the integrity of the market … you're obliged to engage with those that can correct the market."

"This by the way, is the future of the PRI. It needs to pivot to 80% macro stewardship. That would be appropriate, because it is a UN convened institution."

Democratise Savings

"One of the problems I see in capital markets" Catherine Howarth, CEO at ShareAction, told me, "is the underdeveloped democratic rights of pension savers. You don't have the right to elect the people that serve on the board of your pension scheme, nor do you have the right to stand for election to that board."

"Auto-enrolment (AE) has massively increased participation in the system, notably for women, lower paid workers and other disadvantaged groups in the UK labour market."

"But in introducing AE, the government made the decision not to have member nominated trustees."

"A big shortcoming of our pension system, in my view, is that savers have no choice over what workplace scheme they're in, no voice within their scheme, no visibility to their trustees, no representation on scheme boards, and no accountability mechanisms enabling them to question their fiduciaries."

"And yet they bear absolutely all the investment risk. How could anyone possibly describe it as a fit-for-purpose governance regime?"

"If you buy shares in a listed company, you have the right to attend its AGM to hear from the directors about performance and plans they have for the company. You are entitled to pose questions to the board. Why don't we have similar opportunities for pension savers?"

"You might have your entire lifetime's retirement savings in a single scheme, which gives you a far more concentrated risk regarding the performance of that scheme than people would ever be advised to take with underlying investee companies. Yet, there is no opportunity to challenge and query the board members of that scheme."

Nick Robins added, "I'd guess that 90% of the discussions about sustainable finance and within that responsible investment is by suppliers of financial services."

"Involving the users of investment services—their voices and needs—how much people want to be engaged with their pension fund—that still remains a profound issue."

It seems to me that, while most consumer brands have taken steps to build relationships with their "savers" for customers, few pension funds have done likewise.

As part of TCFD reporting, NOW: Pensions, a UK auto-enrolment master trust, undertook focus groups of its savers.

The results were disappointing, but perhaps not surprising. In thinking about responsible investment, there were three conceptual hurdles for savers to overcome.

First, that pensions are invested, many savers considered pensions a form of tax and government spend, a number of savers were not aware that their pension savings were invested in companies.

Second, that investments can be responsible or, the term that resonated most, irresponsible. The two issues that resonated most (with this subsection of savers) were tobacco and guns as irresponsible investments.

Third, that equity investments come with stewardship rights, and that those stewardship rights can be used to engage companies.

For some savers, before these three hurdles, there was surprise that they had a pension at all, with many, particularly, younger, part-time, seasonal or temporary workers unaware of their pension, or perhaps, invested across so many pensions as to have little connection with their individual pension providers. When you have a few hundred dollars in one pension, a few hundred in another and so on, which you cannot imminently access (for younger savers, by as long as 40 years), it's not difficult to understand the apathy.

Advocates of democratising savings say that more democratic rights for pension savers may lead to more understanding of investments and more engagement with sustainability issues.

A related issue is that investment decision-makers and their service providers are not representative of the savers they serve.

The Diversity Project is an initiative "championing a truly diverse and inclusive UK investment and savings industry with the right talent to ... reflect the society we serve and ultimately build more sustainable businesses."

The Diversity Project has established an "asset owner diversity charter", which "formalises a set of actions that asset owners can commit to improve diversity, in all forms, across the investment industry."

The charter includes a questionnaire and a toolkit that asset owners can use to assess the diversity of their asset managers.

The 10,000 Black Interns programme offers paid internship opportunities to those who are Black or of Black heritage, as long as the candidate has a place at a UK university, about to have a place at a UK university, or graduated within the previous 3 years.

So there is some progress, but the financial sector remains undiverse.

I struggle to understand why more NGOs are not seeking to address both these sets of issues: to democratise savings and to diversify investment decision-makers.

NGOs could identify, train and support a more representative pension fund trustee community with expertise on the NGO's topic, across both environmental and social issues: Rather than lobbying existing decision-makers, training the next generation of decision-makers.

"It takes only a very small number of vocal, informed and engaged pension scheme members to be influential" Howarth said.

"A strong parallel is the wider electorate and the political system. It is always a small number of committed citizens that influence and hold accountable their parliamentarians for everyone's benefit."

We need more urgency in addressing representation across the industry and we need to make it easier for savers to voice their sustainability views.

There are other ways we can help democratise savings, including sustainable finance "training" at schools or universities, incorporating sustainability into financial qualifications, improving financial statement communication to savers, and new technologies that allow savers to communicate their sustainability preferences.

Influence

"The issue that motivates me most is designing our policies that would allow the financial sector to maximise its contribution to achieving sustainability goals" Martin Spolc said.

"I started to work on financial services policy to restore financial stability after the banking crisis. I was leading the banking regulation file and we needed to find solutions to the problems that occurred in the financial sector."

"The sustainable finance policy agenda offers the opportunity to the financial sector to be a solution to the problem, this time the problem being climate change and environmental degradation."

"The financial sector has enormous power in supporting companies and economic sectors in their transition efforts and this potential needs to be steered in the right direction."

The way in which the financial sector, and in this case, investors, maximise their contribution to sustainability goals has in recent years preoccupied Martin Spolc and many others.

For me, it comes back to how we think about "influence". Investors themselves are not directly responsible for impact. Rather, they influence others to have impact. But the literature is surprisingly thin on what constitutes effective influence.

A few years ago, WTW's Thinking Ahead Institute began to socialise an idea titled the "4321 pin code". The pin code helps us to start thinking about how we locate investors within an economic and political framework (TAI 2019).

The pin code assumes 10 units of influence or soft power. Four units of soft power can be attributed to governments and policymakers, three units to companies, two to financial institutions and one goes to the individual.

The point of the pin code is to remind us that, as investors, we have 2 in 10 units of soft power, but we can also influence those around us. The numbers themselves are arbitrary, but it helps us to think about investment's role in delivering change, and therefore helps us to focus on what it is we can—tangibly—achieve and how we work with others to achieve it.

"The finance sector's two units, for example, can be multiplied by combining with people (through democratised finance), with corporations (through enlightened ownership) and with policymakers (through empowering regulation)" WTW says (WTW 2021).

As investors, we do not achieve change directly. We influence others to achieve change. And we need to get smarter about thinking about how we use that influence.

Roger Urwin told me, "The 4321 pin code is an opportunity to use systems thinking to solve problems, but it's not yet a well socialised concept."

"Take for example, universal ownership. I was heavily into universal ownership 15 years ago. I wrote a paper about it. That paper was ignored for a decade, but it's now referred to. You can expect ideas to emerge through an S curve, starting out slow, accelerating after a while when the innovators get joined by the early majority thinkers."

"A hyper joined up framework to investing is novel and anything novel takes a decade or two to get traction. People are lured into the benefits of current practice and history, not something from the future where there is not yet any social proof."

"When you think about the Shareholder Commons organisation, TIIP, Jon Lukomnik and others, you get a sense of those plugging away, with good thinking, but not getting a lot of traction. Why? Because it's complicated and a bit too novel."

"The next phase needs to be about all 10 units of power being aligned to the net zero challenge and reaching agreement on policy levers and wider incentives. For the investment industry, this is using its democratised power to engage broad societal support, applying its corporate muscle to engage with the private sector to reduce the destructive effects of business externalities, and using its soft power on government to make progress on the key policy measures."

"Like a price on carbon, clarity on energy priorities, and taxation consistencies. This is about the investment industry taking a systems leadership position to ensure the system can support the future returns needed. You could call it enlightened self-interest."

"As investors, we should have societal support through our savers and the democratisation of savings. We can leverage corporate muscle because we own these companies. We can influence and mobilising policy and regulation through macro-engagement."

"I've found it interesting how net zero is ultimately some form of partnership through the ecosystem."

"At the state level, it's the nationally determined contributions. For companies and investors, it's net zero commitments. Companies and investors can push the state and the state can push companies and investors."

"The 4321 pin code is nothing more than expressing this relationship succinctly. It's not innate thought leadership. It's just a mechanism for communication."

With leadership from colleagues at the team, notably deputy CIO, Keith Guthrie, we tried to better understand influence at Cardano, the UK and Dutch fiduciary investor, asset manager and advisor where I previously worked.

In 2022, Cardano published its model of influence, which is practical and implementable, setting out the activities an investor can undertake, how the activities have influence and how we should think about these activities in relation to sustainability goals.

The first part is a clear articulation of two simultaneous objectives: to maximise risk-adjusted returns and to maximise real-world sustainability impact. Institutional investors increasingly share this "double materiality" view and want to incorporate both lenses into their investment process. Yet there is little clarity about what is meant by influence, real-world sustainability impact and how to measure it.

The second part is tiers of activities undertaken by investors. The activities are tiered by the extent of their influence, because some of the activities are more impactful. The tiers are not mutually exclusive, they're intended to be used in conjunction.

The first tier includes:

- The supply of new capital, debt or equity to a company or government, where this has an environmental or social objective (such as a primary issuance green bond).
- Collaborative company engagement on sustainability topics, as well as co-filing resolutions on sustainability topics.
- Public policy engagement on sustainability topics. Public policy engagement is a natural extension of stewardship. Policy change is often the result of multiple influencing factors, but when it does happen, it has very high impact.
- New and innovative forms of impact investment.

The second tier is engagement with companies on ESG issues and voting on shareholder resolutions, director appointments and other resolutions.

Individual engagement is less impactful than collaborative engagement, but does have impact, particularly on topics where there is not yet industry consensus, and there is no collaborative initiative, or no pool of investors considering similar actions. Engagement should be across asset classes.

Voting is foundational. For many equity investors, voting is expected by their clients. But voting can be used as a form of escalation, in particular, votes against the chair, members of the board or remuneration packages.

The third tier is cost of capital influence. All things being equal, buying shares pushes up the price of the shares, and therefore reduces the cost of capital for the company, enabling the company to grow (disproportionately to their competitors). The mechanism is indirect, as markets are complex, and pricing is subject to multiple factors. As such, it's difficult, almost impossible, to determine whether you're having a particular influence.

Tier three is most powerful when it is directional—in other words, when investors are moving in the same direction. The early movers will hope to benefit in investment returns (their reward for anticipating the direction). Tier three is most effective when it is in the mandate, the contractual relationship between asset owner and asset manager.

The fourth tier includes activities that are not influential. This includes macro strategies or trend following strategies.

The purpose of the model of influence is to clarify the actions the investor can undertake—and provide a framework to support the investor understand each lever's degree of influence and how the action contributes to real-world impact.

Investors may disagree with the tiering, but thinking through the range of activities, whether the activity is direct or indirect, the degree of influence, and how different activities can be combined to maximise influence is, I think, worthwhile, and transversal. The framework also helps us to think about escalation.

Politics, Stupid

"One interviewee told me that "part of looking after other people's money involves looking after other people and well-being."

I have personal experience of politics. In 2015 I stood for Parliament as the Labour Party's candidate in Battersea, a marginal constituency in South West London. Battersea is one of the country's most unequal constituencies. Clapham Junction, a major London train station, divides the constituency. Those North of Clapham Junction have a life expectancy 10 years lower than those South of Clapham Junction.

Being a parliamentary candidate wasn't an enjoyable experience. It was of course unpaid. I took three months unpaid leave. That was tough at the time. I would imagine the energy and time I spent on the campaign put my career back three years, perhaps more.

And I was lucky—my employer (the PRI, and in particular, PRI's CEO Fiona Reynolds) was supportive. For some, the decision to stand—regardless of whether or not you win—is career-ending. Who'd employ a former MP? Who'd employ a former (losing) MP candidate?

During the campaign, we moved our daughter, who was 18 months old at the time of the election, away from our terraced home's street-facing window for fear of a brick coming through. An envelope addressed to me containing cat litter was posted through our campaign office letter box. The opposition did a land registry check on my house to find out my mortgage.

The Economist (as in, The "actual" Economist magazine) threatened legal action because a campaign leaflet, they said, resembled their magazine. Volunteer lawyers supporting the Labour Party prepared our response. But it was a nervous few weeks. "Banter" my campaign lead texted me when we received the letter from The Economist's lawyers. It didn't feel like banter. I ended my subscription.

A Labour Party member called me a fucking prick banker on twitter. The Evening Standard said that I was "seeking a left-turn" and that I was "keen to reinvent" myself as a "left-winger" saying that I'd worked for Oxfam "assisting with its anti-capitalist campaigning".

Part-way through the campaign, and less than a month after our daughter was born prematurely, not feeding, and having spent her first two weeks in hospital, the Sunday Times ran the headline "Labour candidates tell Miliband to 'hug a banker'" having secretly recorded an event I was speaking at. The journalist selectively quoted from a 90-minute discussion.

I remember it as a high quality, substantive and interesting event.

Left-leaning Guardian journalist Polly Toynbee, who I'd met and previously held in high regard, joined the critique. She didn't ask to speak to me. Just wrote about me. It wasn't fun.

During my two-year candidacy, I met some wonderful people and I think I made a difference—volunteering at food banks, attending Christmas Day events for older people, running the marathon for the local hospice—but I didn't enjoy it.

In 2015 I lost. The election was on a Thursday. On the Saturday I took my daughter to the local swimming pool. A few people recognised me and said hi, offering their commiserations. On Monday I went back to work. Ed Miliband, the then Labour leader, didn't write to me, nor did his office. The Labour Party head office didn't write to me. No note of thanks. Nothing. Nada.

In 2017, I decided not to run. Inevitably, the seat went Labour. It was a tough time personally. We had moved to Paris. Being a Member of Parliament is perhaps the greatest of privileges. But I also think, in some respects, I dodged a bullet.

Whenever I get just an inkling to re-stand a good friend of mine writes to me. "Don't do it." "Want to be a good dad?" "Don't do it." What a tragedy of our political system. Parent or politician? You can't do both.

MPs are subject to 24-hour-a-day scrutiny. Can't take a holiday. Compared to professionals of similar standing, low pay. Anti-social hours. For those in marginal seats, no job security. Constant online abuse (far worse for women, and black and ethnic minorities, than it was for me). No support.

During the 2015 campaign, I spent time with Jo Cox. We had both worked at Oxfam. We were a similar age. We would sit together at the back of Labour Party briefings. On 16 June 2016, Cox, then MP for Batley and Spen was repeatedly shot and stabbed by a 52-year-old man. Cox was 41, with two small children. She was a remarkable woman. Not much has changed to improve the security of—and respect towards—MPs since.

Society has this counterproductive way of benchmarking MPs on "localness". MPs are not mayors. Their job is to legislate. And yet, we want MPs that "come from" their constituency, opening school fetes, cleaning up the local brook. Few swing voters would reward an MP for spending a week scrutinising complex legislation on topics such as responsible investment. No. Opening a new shop or some-such for a Facebook photo shoot. Yes.

Frankly, we get the politicians we deserve. Or as journalist and author Isabel Hardman puts it, "Why we get the wrong politicians." The number one requirement? Money. Number two requirement? A thick skin. Even then, I still have tremendous respect for the MPs we end up with.

But how's this relevant to responsible investment? Well, at the end of day, investment portfolios are a mirror to the real economy. The sustainability of an investment portfolio, plus or minus, is a direct function of the sustainability of the economy into which we invest. And politicians are responsible for the rules of the game.

One of the most effective actions an investor can take is to achieve real-economy policy change (or to use another term, systemic stewardship). This is precisely why companies pay arm and leg for the very best lobbyists, join a host of expensive lobby groups (that trounce the fee paid to PRI or IIGCC) and send their executives to high fee dinners to meet with politicians. Or if you're in the US, skip the middleman and simply write a cheque.

Real-economy policy change—or, to use a simpler word, politics—is how we achieve a more sustainable financial system. So why then, are responsible investors silent on politics?

There are a few reasons.

1. A perennial reason of not wanting to alienate a saver or client base.
2. A sense of "trust the process". Politicians are reasonable people, let's hope they will make the right decisions.
3. The benefits of big incumbency (think about post-Brexit Nissan in Swindon), which will get you advantageous treatment with respect to tax, labour rights and trade.
4. The benefits of privileged access to government.

5. Many companies, and certainly many investors, simply lack the skill set to engage policy makers. For me, this is made worse by investors' relative comfort commenting on overseas politics through sovereign bond engagement.

It is changing. As investors pursue higher impact strategies, some investors are considering responsible political engagement.

In 2022, writing an op ed for the trade publication, Environmental Finance, I argued that if responsible investment is to matter, it is inherently political (Environmental Finance 2022). The prompt for the article was the Taxonomy and the decision by EU policymakers to badge (in certain instances) gas and nuclear as green.

"The EU Taxonomy complementary delegated act is only the start" I said. "We've seen this in the US in slow motion. Successive administrations have flip-flopped on Department of Labor ESG integration requirements."

"The EU Taxonomy is a step change because the worlds of financial, industrial and fiscal policy collide. The Taxonomy acts as a mirror to the EU's 'Fit for 55' strategy. For the first time, European policymakers are looking in that mirror. Who wins and who loses is political."

"In recent years, the sustainable finance community has become more comfortable with policy engagement—albeit with technical policymakers and ministries on policy design, rather than politics. The delegated act represents an inflection point for sustainable finance. The Taxonomy is just the start."

"Whether we like it or not, the future of sustainable finance is inherently political."

For investors, this should start by elevating policy engagement as part of the engagement strategy.

Few investors have a "head of policy" whose job it is to engage policymakers on sustainability topics. Mostly, the head of policy is reactive, responding to sustainability-related financial regulation and providing interpretation, rather than proactive policy engagement on real-economy topics, such as energy policy, buildings efficiency, transport, agricultural policy or carbon sequestration projects—or even, a carbon tax.

Investors have influence with politicians for a few reasons. Investors have democratic legitimacy and are "actors in the real economy" investing on behalf of millions of savers. Investors also have influence because the way in which they direct capital has real-world consequences, and investors have expertise. Investors can support policymakers in achieving their goal.

In 2014, I co-authored the PRI report, the Case for Investor Engagement in Public Policy. Unfortunately, almost a decade later, it remains relevant.

Policy engagement to pursue sustainability outcomes is not yet standard practice. But for me, it's a natural extension of stewardship and part of the future of responsible investment.

I asked Alex Edmans whether he believes investors should engage policymakers. "Yes, because if you're doing it through the government, then that's fine. I think it's powerful for investors to do that, saying to governments that our own clients care not just about their retirement income, but also about whether the planet is 2 degrees Centigrade warmer."

References

Environmental Finance (2022), The Future of Sustainable Finance Is Political. [online]. Available from: https://www.environmental-finance.com/content/analysis/the-future-of-sustainable-finance-is-political.html (Accessed, February 2023).

Forbes (2022), The Role of Systemic Stewardship in Addressing Income Inequality. [online]. Available from: https://www.forbes.com/sites/bobeccles/2022/01/27/the-role-of-systemic-stewardship-in-addressing-income-inequality/ (Accessed, May 2023).

Forum for the Future (2019), What Is Systems Change? [online]. Available from: https://www.forumforthefuture.org/faqs/what-is-system-change (Accessed, January 2023).

Aviva Investors (2022), Macro Stewardship. [online]. Available from: https://www.avivainvestors.com/en-gb/views/aiq-short-reads/2022/09/macro-stewardship/ (Accessed, February 2023).

Hawley and Lukomnik (2021), Moving Beyond Modern Portfolio Theory: Investing that matters, Routledge.

LocalGov (2022), Gove Urges Pension Funds to Add to Russian Pressure. [online]. Available from: https://www.localgov.co.uk/Gove-urges-pension-funds-to-add-to-Russian-pressure/53866 (Accessed, February 2023).

Lukomnik (2022), Moving Beyond Modern Portfolio Theory [online]. Available from: https://www.youtube.com/watch?v=tSR0vgOAOtg (Accessed, January 2023).

Markowitz (1952), Portfolio Selection [online]. Available from: https://onlinelibrary.wiley.com/doi/abs/10.1111/j.1540-6261.1952.tb01525.x (Accessed, January 2023).

PRI (2017a), The SDG Investment Case. [online]. Available from: https://www.unpri.org/download?ac=6246 (Accessed, May 2023).

PRI (2017b), Sustainable Financial System. [online]. Available from: https://www.unpri.org/sustainability-issues/sustainable-markets/sustainable-financial-system (Accessed, January 2023).

TAI (2019), The 4-3-2-1 PIN Code for a More Sustainable Economy. [online]. Available from: https://www.thinkingaheadinstitute.org/news/article/the-4-3-2-1-pin-code-for-a-more-sustainable-economy/ (Accessed, January 2023).

TAI (2020), Top 100 Asset Owners | The Most Influential Capital on the Planet. [online]. Available from: https://www.thinkingaheadinstitute.org/content/uploads/2020/11/TAI_AO100_2020.pdf (Accessed, May 2023).

TIIP (2022), Systemic Stewardship: Investing to Address Income Inequality. [online]. Available from: https://tiiproject.com/wp-content/uploads/2022/01/TIIP-Stewardship-Final.pdf (Accessed, February 2023).

Universal Owner Initiatives (2020), What Is Universal Ownership Theory. [online]. Available from: https://www.universalowner.org/universalownership theory (Accessed, May 2023).

Urwin, Roger (2011), Pension Funds as Universal Owners: Opportunity Beckons and Leadership Calls. [online]. Available from: https://papers.ssrn.com/sol3/papers.cfm?abstract_id=1829271 (Accessed, January 2023).

WTW (2021), The Investment Industry Should Unleash Its Fiduciary Capital to Address Climate Change. [online]. Available from: https://www.wtwco.com/en-GB/Insights/2021/10/the-investment-industry-should-unleash-its-fiduciary-capital-to-address-climate-change (Accessed, January 2023).

16

Final Words

In Their Own Words

I asked those I interviewed for this book what next for responsible investment. Here is a selection of their responses.

Bob Eccles told me: "The standards are going to be important. The political winds will die in the US."

"ESG integration will be better done across all investing. I'd like to see the term ESG integration disappear. We'll simply have investing. Everybody will be doing it."

"More movement in the direction of universal ownership, but it's hard to predict how that will evolve."

"We'll see clarity about naming. Distinguishing between ESG integration and impact, you'll see more clarity around that. Is it concessionary, is it non-concessionary?"

"I don't really know where GIIN ends up in this. They're in such in a niche place. PRI will exist. It will have to be country or territory specific. The metric of success was number of signatories and AUM. I think that's less relevant now. I think its role is more about stewardship and public policy. That's a big opportunity for the PRI. On the investor side, the Gorilla is the PRI."

Fiona Reynolds told me, "I would like us to never again have to talk about responsible investment, it should just be investment. We shouldn't need to have all of these different terms. We should just allocate money efficiently and in a way that respects people, doesn't exploit people or planet, and is

© The Author(s), under exclusive license to Springer Nature
Switzerland AG 2023
W. Martindale, *Responsible Investment*, https://doi.org/10.1007/978-3-031-44536-1_16

within planetary boundaries. That's what I want to happen. It's what I have always wanted to happen, and why I work in this space."

Erinch Sahan said "I think there will soon come a reckoning when it's going to become really clear that the win wins have been done to death. We'll be exhausted by trying to find a way to put old wine in new bottles. Responsible investment will lose credibility as a result."

"Over 10 years ago at Oxfam, we published Better Returns in a Better World. To engage the finance sector, it was tactically useful. But that mantra now feels outdated. Instead, fair returns in a thriving world is probably the update to that mantra."

"Addressing this is going to put the onus on regulatory reform and it's going to require some form of international collaboration to ensure that the regulated countries don't push away capital. The transformation will be hard, but I can't see how we can avoid it."

Richard Roberts said, "I feel like the responsible investment movement has lost its way."

"If you go back to the founding of the movement 15–20 years ago, it peaked around 2019 / 2020, with a massive conflation of ESG, responsibility, sustainability and impact. There was this narrative of, 'mainstream investors are embracing ESG, that equals sustainability impact'. There's been a healthy disintegration of that over the last 2–3 years."

"If you look at some of the PRI principles around integration and disclosure then I don't see that as part of responsible investment today. Rather, this is just good strong risk management that all investors should practice."

"So if that's no longer responsible (partly because of the success of PRI in mainstreaming ESG), there's definitely an opportunity for responsible investment to move into a different space which is more about impact and potentially about ethics. The next frontier for responsible investment has to be much more focused on impact."

Nathan Fabian said, "There's no doubt in my mind that both capacity and willingness to calibrate and act on investment decisions relative to planetary boundaries and international rights frameworks is the future of responsible investment."

Philippe Zaouati said, "The future of responsible investment is at a crossroads. Everyone is speaking about ESG issues and sustainable finance today, so it's not a niche anymore, but the centre of how we think about investment. And yet, if we look at what's happening in the US, it's becoming a political issue and very polarized."

"In terms of what the future of responsible investment should be, I think we still need a part of the market that is ahead of the curve on sustainability issues. While I like mainstreaming, that doesn't mean we don't need pioneers from a number of very committed players."

"The work underway to change the economy is still work in progress, and we've got a lot to do, and to do it, we need investors that have deep conviction on sustainability issues. There's a risk that mainstreaming will lead to a dilution."

"I have this discussion with my shareholders. The responsible investment market is huge. Mirova could become a big player, with less committed products, and get access to a wider market. But for me, we need to remain high conviction, ahead of the curve, pioneering new frontiers for responsible investment."

Nick Robins said, "To some extent, maybe the future is separating the things that are now doable from the things that remain really hard. For example, the energy transition, because of the huge technological benefits we now have, is completely doable: whether it is delivered on time is another matter and has much to do with the power of incumbent fossil fuel interests in business, finance and politics."

"But there are issues which are much more problematic, such as inequality, where less progress has been made, not just by investors, but society at large. The shift in climate policy to the green stimulus approach embodied by the Inflation Reduction Act is a sign of policymakers trying to connect the energy transition with job creation in economic heartlands, a kind of just transition if you like."

Catherine Howarth said, "For me, the big thing that no one's talking about is the need for democratic reform in capital markets. The industry itself is uninterested in pension savers having more agency, more voice and in being empowered to hold the people in charge of investing their capital to account."

"Responsible investment is dominated by intelligent people but most of them come from the elite in our society. And it gets very technical at times, which can be alienating for underlying savers."

"The millions of people whose money it is are outside the conversation about responsible investment even though many of them are directly affected by or care deeply about the issues that we work on in responsible investment. I don't think we'll get profound change until we address that democratic deficit in capital markets."

"I hope and believe that savers are beginning to clock that they have a stake in the system, and should push for change until their voices and priorities are formally represented. It's democracy in pensions that would create meaningful change in the type of conversation and the scope of issues that are covered. Many members are not engaged. This undermines and weakens the good work underway in responsible investment."

And Howarth's advice for responsible investment professionals, "My advice is to work for your firm but think of yourself as a change agent in the wider system. Responsible investment is not yet delivering against its promise. We can change that by dreaming big together."

Steve Waygood said, "I want the future of responsible investment to be a challenging environment for those that are greenwashing. I want regulators and civil society to step up and challenge those that are making illegitimate claims. To attack the laggards, not just those with big brands. If you're an NGO, big brands will get you more copy. But, in general, they are often not the main laggards."

"Green hushing will become a big problem in US. Under Ursula von der Leyen, responsible investment has a very bright future. There's lots of progress to be made."

"A lack of sustainability is the world's biggest market failure."

"[Responsible investment can] save finance from itself. In short, finance can save the world. For its own interests, as well as the world."

[But] "there's an enormous amount of noise now in responsible investment, and not much signal to achieve real-world sustainability impact. Responsible investment professionals should tune in carefully to the signal and not just amplify noise. Ask yourself: 'Is this thing I'm doing going to make a dent on the seismic sustainability issues that will harm economic growth?'. If not, focus on something else that will."

Martin Spolc said, "The sustainable finance policy agenda is about integrating sustainability considerations in the way how the financial sector operates and maximising the positive impact it can bring on climate, nature and people. This gives us a better understanding that there are planetary boundaries and we cannot exploit natural resources indefinitely."

"This had not been reflected sufficiently in economic discussions and our financial services policies in the past."

"I am a financial analyst myself and I was educated on the premise that everything is risk and return—a two-dimensional scheme. Of course, sustainability will eventually, one way or another, affect risk and return."

"But the consideration of impact of investments on climate, nature and people in the equation was largely missing."

"When my children ask me what I am doing, I say, 'look, currently banks and investors are putting money into projects that will ultimately lead to a climate change of over 3 degrees Centigrade. Something needs to change to improve the situation."

"We need to get the financial sector funding projects that will improve the impact on climate, environment and people."

"To ensure that over time the financial sector is financing projects that are compatible with our future. To achieve change, it is important to mainstream these considerations in the regulatory framework."

Here's My Take

Responsible investment has lost its way. For all the noise it is unproven. In its current form, its contribution to real-world sustainability impact is limited. This is despite the time, costs and complexity involved in much of the new disclosure requirements.

ESG initiatives are not free. Someone somewhere is paying for it and in the case of pension funds, the costs are often borne by working people.

Mechanical (and often unthinking) integration of black-box ESG scores in investment decision-making, with reporting on often arbitrary thresholds, does not make the investment responsible. That's not to say ESG data sets should be ignored. Good risk management of unpriced ESG risks is prudent investment. But any real-world impact is a by-product.

To be responsible, an investor must understand real-world impact, and use influence to maximise positive real-world impact and minimise negative real-world impact through investment and stewardship, with progress calibrated to international frameworks.

The most effective responsible investment strategies commit to a change objective and undertake actions to make progress towards that change objective, with accessible reporting in place for clients and savers.

On climate change, we need to accept that divestment is unlikely to influence oil and gas companies. High conviction stewardship and investments in green solutions are more useful tools.

The most effective responsible investors understand the system into which they invest, how that system may work for or against real-world sustainability impact, and understand how to maximise their use of influence to drive change through the actions available to them as an investor.

The motivation can and in most cases, often should, be financial, even where the investor pursues a real-world impact objective. For many investors, their or their client's financial performance is subject to the performance of the real economy, where sustainability issues represent risks to the system.

Responsible investors should also establish minimum ethical standards. There are some investments that are simply inconsistent with acting responsibly. Some investors may go further, either because that is the standard to which they hold themselves or respond to client demand.

Finally, responsible investment should understand its limitations. There may be issues that require policy change where investors are not well-placed (or uniquely unable) to pursue that change. If so, this is an issue where responsible investment should sit out, however egregious.

There's more than enough for us to be getting on with where we can make a difference.

What Are CDOs?

CDOs stand for collateralised debt obligations.

To understand CDOs we have to first understand CDSs.

CDS stands for credit default swap. A CDS is a derivative on a bond (it derives its price from a bond).

A CDS has a notional and a coupon just like a bond. The buyer (investor) receives an annual payment (the coupon). If the company (and therefore, its bond) defaults, the investor pays the notional. It is a bit like a form of insurance. Whereas, for a bond, the notional is paid upfront, for a CDS, it's paid on default of the underlying company.

A CDO is a portfolio of CDSs. The investor's risk is determined by what are called subordination levels, specific to the CDO. The lower the subordination, the higher the default risk, the higher the coupon.

To be investable, the issuer of the CDO, an investment bank, would work with a credit ratings agency to give the CDO a rating, typically an investment grade rating, BBB—or above.

Most of the CDOs I worked with were synthetic. They didn't actually exist, but rather, were fully modelled.

If it sounds unbelievable, that's because it is. Our collective ability to over-complicate, over-engineer, over-legalise, and in this case, over-securitise our day-to-day lives is extraordinary.

Bringing purpose back to the financial system is long overdue. Responsible investment is absolutely part of that.

With indebted governments in the aftermath of the global financial crisis and pandemic, rising inequality, rising populism, the growing severity of climate change and the overshoot of a range of other planetary boundaries, the extent to which private capital can contribute and should contribute towards societal goals is one of the defining issues of our generation.

The story of responsible investment is one that's still underway.

As we consider its future, there are a few ways to anchor our thinking about responsible investment.

- One is greatest good for greatest number (this is economic orthodoxy, but perhaps responsible investment provides us with a lens to really challenge whether this is how capital markets are working).
- One is most fortunate help the least fortunate, which in a way, aligns with Rawls, who argued that a just society should be structured so that the least advantaged members are as well off as possible.
- One is natural social concern—seeking what is good for others as well as for ourselves.
- One is Kant—categorical imperative—it's the right thing to do.

Perhaps responsible investment is all of the above. But for me personally, it's Kant. It's the right thing to do. That we, as humans, have intrinsic, non-financial reasons for wanting to do things, even if the financial constructs through which we work have only extrinsic ones.

We should think more about whether responsible investment as its currently practised is working the way we expect it to, whether there's more we can do to maximise positive real-world sustainability impact and minimise negative real-world impact, and if so, well, I think we have a categorical imperative to do so.

As Kim Robinson writes in Ministry for the Future, "it's easier to imagine the end of the world than the end of capitalism".

For now at least, responsible investment remains a worthwhile pursuit.

Reference

Robinson (2020), The Ministry for the Future. Orbit Books.

Glossary

AOA Net Zero Asset Owners Alliance
bp Basis point
BP British Petroleum
CA100 Climate Action 100+
CBAM Carbon Border Adjustment Mechanism
CBI Climate Bonds Initiative
CDO Collateralised Debt Obligation
CDP Carbon Disclosure Project
CDS Credit Default Swap
COP Conference of Parties
CRISA Code for Responsible Investment in South Africa
CSR Corporate Social Responsibility
DOL US Department of Labor
DWP UK Department for Work and Pensions
EC European Commission
EFRAG European Financial Reporting Advisory Group
ERISA US Employee Retirement Security Act
ESG You know that one. Some say, the G is corporate governance (I prefer, just, governance)
ESMA European Securities and Markets Authority
ETF Exchange Traded Fund
ETS Emissions Trading Scheme
EU European Union
EVIC Enterprise Value Including Cash
FCA UK Financial Conduct Authority

© The Editor(s) (if applicable) and The Author(s), under exclusive
license to Springer Nature Switzerland AG 2023
V. Martindale, *Responsible Investment*, https://doi.org/10.1007/978-3-031-44536-1

FFD Financing For Development
FRC UK Financial Reporting Council
FSA Japan Financial Services Agency
GFANZ Global Financial Alliance for Net Zero
GHG Greenhouse gas
GIIN Global Impact Investing Network
GP (private equity) General Partner
GP UN Guiding Principles on Business and Human Rights
GRI Global Reporting Initiative
GSG Global Steering Group
HLEG EU High Level Expert Group
ICGN International Corporate Governance Network
IEA International Energy Agency
IFRS International Financial Reporting Standard
IGCC Investor Group on Climate Change
IIGCC Institutional Investor Group on Climate Change
IIRC International Integrated Reporting Council
ILO International Labor Organisation
IMP Impact Management Project
IORP EU Institutions for Occupational Retirement Provision
IPDD Investor Policy Dialogue on Deforestation
IPR Inevitable Policy Response
IPSF International Platform on Sustainable Finance
ISSB International Sustainability Standards Board
MMMM Make My Money Matter
MNE Multinational Enterprise guidelines
MPT Modern Portfolio Theory
NDC Nationally Determined Contribution
NFRD EU Non-Financial Reporting Directive
NGO Non-governmental organisation (a charity)
NZAMI Net Zero Asset Managers Initiative
NZBA Net-Zero Banking Alliance
NZICI Net Zero Investment Consultants Initiative
PAB Paris Aligned Benchmark
PAI Principal Adverse Indicator
PAII Paris Aligned Investment Initiative
PBAF Partnership for Biodiversity Accounting Financials
PCAF Partnership for Climate Accounting Financials
PCRIG Pensions Climate Risk Industry Group
PRB Principles for Responsible Banking
PRI Principles for Responsible Investment
RBC Responsible Business Conduct
RI Responsible investment
SASB Sustainability Accounting Standards Board

SBTi Science-Based Targets Initiative

SDG Sustainable Development Goal

SDR UK Sustainability Disclosure Requirements

SEC US Securities and Exchange Commission

SFDR EU Sustainable Finance Disclosure Regulation

SRD EU Shareholder Rights Directive

SRI Socially responsible investment (typically considered as ethical investment). Sometimes SRI is sustainable and responsible investment.

TCFD Task Force on Climate-related Financial Disclosures

TEG EU Technical Expert Group

TNFD Task Force on Nature-related Financial Disclosures

TPR UK The Pensions Regulator

UKSIF UK Sustainable Investment and Finance Association

UN PRI See PRI

UNEP FI United Nations Environment Programme Finance Initiative

WACI Weighted Average Carbon Intensity

WBA World Benchmarking Alliance

WEF World Economic Forum

Index

Printed by Printforce, United Kingdom